T4-BBU-885

Cambridge Studies in Philosophy and Biology

General Editor
Michael Ruse   University of Guelph

Advisory Board
Michael Donoghue   Harvard University
Jean Gayon   University of Paris
Jonathan Hodge   University of Leeds
Jane Maienschein   Arizona State University
Jesus Mosterin   University of Barcelona
Elliott Sober   University of Wisconsin

Alfred I. Tauber   *The Immune Self: Theory or Metaphor?*
Elliott Sober   *From a Biological Point of View*
Robert Brandon   *Concepts and Methods in Evolutionary Biology*
Peter Godfrey-Smith   *Complexity and the Function of Mind in Nature*
William A. Rottschaefer   *The Biology and Psychology of Moral Agency*
Sahotra Sarkar   *Genetics and Reductionism*
Jean Gayon   *Darwinism's Struggle for Survival*
Jane Maienschein and Michael Ruse (eds.)   *Biology and the Foundation of Ethics*
Jack Wilson   *Biological Individuality*

Biology and Epistemology

Edited by

RICHARD CREATH
Arizona State University

JANE MAIENSCHEIN
Arizona State University

CAMBRIDGE
UNIVERSITY PRESS

LIBRARY
UNIVERSITY OF ST. FRANCIS
JOLIET, ILLINOIS

PUBLISHED BY THE PRESS SYNDICATE OF THE UNIVERSITY OF CAMBRIDGE
The Pitt Building, Trumpington Street, Cambridge, United Kingdom

CAMBRIDGE UNIVERSITY PRESS
The Edinburgh Building, Cambridge CB2 2RU, UK   http://www.cup.cam.ac.uk
40 West 20th Street, New York, NY 10011-4211, USA   http://www.cup.org
10 Stamford Road, Oakleigh, Melbourne 3166, Australia
Ruiz de Alarcón 13, 28014 Madrid, Spain

© Cambridge University Press 2000

This book is in copyright. Subject to statutory exception
and to the provisions of relevant collective licensing agreements,
no reproduction of any part may take place without
the written permission of Cambridge University Press.

First published 2000

Printed in the United States of America

Typeface Palatino 10/13 pt.   *System* MagnaType™ [AG]

*A catalog record for this book is available from
the British Library.*

Library of Congress Cataloging-in-Publication Data
Biology and epistemology / edited by Richard Creath, Jane Maienschein
p.   cm. – (Cambridge studies in philosophy and biology)
ISBN 0-521-59290-9 (hb). – ISBN 0-521-59701-3 (pbk.)
1. Biology – Philosophy. I. Creath, Richard. II. Maienschein,
Jane. III. Series.
QH331.B55   1999
570'.1 – dc21   99-22990

ISBN 0 521 59290 9 hardback
ISBN 0 521 59701 3 paperback

574.01
C912

$18.65

Pub.

5-2-00

Contents

vii

Introduction

RICHARD CREATH AND JANE MAIENSCHEIN

Epistemological issues have always been at the heart of philosophy of science. After all, science considered as a product is a set of organized knowledge claims. Considered as a process or institution, science is the (more or less) organized attempt to devise and defend – that is, to justify – such claims to know the world around us. Just as scientists find and refine their evidence and articulate their arguments so that this bit of data bears on that bit of theorizing, so philosophers have sought general accounts of what scientific observations and arguments can be. Taken together, these accounts comprise a theory of scientific knowledge. And while it may not be the only way to look at science, the attempt to say what makes it *science* is bound to remain a central concern.

Within the past two decades or so, philosophy of biology has emerged as an important and recognized specialty within philosophy of science. For the first time, large numbers of scientifically well-informed philosophers (often joined by historians and reflective biologists) have systematically examined the vast domain of biological work. Are biological claims to knowledge justified in some *different* way (perhaps because those claims are historical, or about life, or about systems unmanageably complex, or perhaps because we have an emotional or political interest in the outcome)? Or is biology fundamentally similar to nonbiological sciences, with similarities that would have gone undetected but for the examination of biology on its own terms?

For the most part, the studies that have appeared so far in the philosophy of biology have concentrated on the character of biological theories (especially evolution) and concepts (e.g., species and individuals as units of selection). As interesting as these interpretive, semantic, and ontological questions are, they do not focus on specifically epistemic concerns. There have, of course, been volumes written about so-called evolutionary epistemology. But these have sought the structure of an answer to the question "why do some theories survive?" in the structure of evolutionary theory. However interesting, this is a long way from answering the epistemic questions about evidence and argument and a long way from the consideration of the full range of biological sciences.

Certainly there has been epistemological work about biology, and from a rich variety of sources, too. What has not appeared is a single volume that focuses on biology and epistemology, that brings together and into focus epistemological work on the full range of domains within biology, and that likewise brings together work focused on biology that illuminates a range of epistemological issues. This volume places the individual efforts within the context of the general and common concerns, and we believe that the result will be of value to philosophers and historians of science, to biologists, and even to those interested in some aspects of science policy.

The volume is organized into three sections, though there are many overlaps and a number of the chapters could have fallen into more than one section. The first focuses on a central idea of the nineteenth century: evolution and its contemporary philosophy of science. What view did Darwin and leading evolutionists hold concerning the nature of evidence and its relation to theory? What was the relation of philosophy of science to biology? *Michael Ruse* considers "Darwin and the Philosophers," by which he means primarily John Herschel and William Whewell. Philosophical ideas played an important role for Darwin, Ruse contends, and we can even see Darwin's work as an attempt to respond both to Herschel's British empiricist demand for knowledge of *verae causae* through direct experience and to Whewell's rationalist understanding of those same *verae causae* as accessible through a process of consilience of inductions. Indeed, Ruse sees Darwin as constructing his "one long argument" to build a theory with *verae causae* at its very heart. Darwin's efforts to take the middle ground and to satisfy both sets of criteria left him satisfying neither of the philosophers of science he sought to follow. In the "twilight" of their respective careers, Herschel could not accept natural selection as the mechanism for evolutionary change, and Whewell rejected the enterprise altogether. Even younger men more sympathetic to the evolutionary view, like Thomas Henry Huxley, remained unimpressed by the rationalist elements of the approach. In order to understand the reception of Darwin's ideas, Ruse argues, we need to understand the interplay of alternative contemporary philosophical ideas.

Jon Hodge denies that Whewell had a significant impact on Darwin's own standards of scientific acceptance and argumentation. Even from the *Beagle* years Darwin was a disciple of Lyell and Herschel, holding that a science must explain by reference to *verae causae* – that is, causes for which there is evidence independent of the facts they are invoked to explain. By the summer of 1838 Darwin held the commonplace views that it is a virtue to connect disparate phenomena as well as to allow successful prediction and that purely hypothetical conjectures should be replaced where possible. At the same time he planned a book which would separate evidence for the theory from its explanatory use in unifying many different facts.

Darwin then carefully studied Whewell's *History of the Inductive Sciences* and assimilated some views on the a priori. But Whewell made no dent in Darwin's methodological loyalties to Lyell. Certainly Darwin did not learn then about Whewell's proposal for an alternative to the *verae causae* ideal, the consilience of inductions, which appeared only in 1840 after Darwin's theory of natural selection and the arguments for it were already firmly in place. The consilience idea denies a distinction between evidence for a theory and its explanatory use, a distinction that Darwin continued to maintain. According to the consilience idea we get strong verification when (a) the theory explains many different facts and (b) the theory explains many facts unknown when it was first conceived. The first condition is not new, having been defended by Herschel and even Darwin. The second is new, but Darwin must have ignored it, for he never discussed the relevant issues. This was true of Darwin's writing on through the *Origin*. Thereafter he sometimes backpedaled a bit on the *verae causae* ideal, suggesting that it may be too demanding and putting more emphasis on explanatory unification. But Darwin never fully repudiated *verae causae*; Whewell was not his inspiration on explanatory unification; and he never addressed the concerns of (b).

In order to explore the larger question of what effects philosophy of science and science have on each other, *David Hull* also focuses on the case of Darwin and his philosophers. He thus tackles a set of questions similar to Ruse's, but introduces John Stuart Mill into the mix. Mill, as the first truly inductivist philosopher of science, was more representative of the time, when both the "hypothetical" and "deductive" cores of Herschel's and Whewell's philosophies remained suspect. Mill's *System of Logic* provided a method for science, and according to Mill's interpretation of the strict logical standards, Darwin had provided a theory that was logical and that could be true. It could be. But, Mill believed, it was not. A logical and possibly true theory was not a proven theory, and Darwin did not have proof. Instead, a theory based on intelligent design was more defensible for Mill. Thus, the fact that Mill endorsed Darwin's method of constructing a theory did not help Darwin to establish the validity of, or to justify the belief in, evolution by natural selection. What does this tell us about the relation of philosophy to science? Well, Herschel, Whewell, and Mill all rejected Darwin's theories, even though Darwin sought to base his views on their philosophies. Surely, Hull concludes, this calls into question whether philosophy of science and science really help each other very much.

Robert Richards steps back and suggests that most scholars to date have adopted an unacceptable essentialist view of scientific theories. Instead he offers an historical approach, where individual theories are individuals with developmental histories and lineages to which they belong. Thus, any study of theories should proceed historically, and with full awareness of the changeability of individuals over time. In particular, he cites study of Darwin as a problem. Scholars have taken Darwin's theory of evolution by natural selection as if it were one thing, and they have sought to discover its origin and to consider its reception. In particular, they have found it important to deny any progressivist thinking on Darwin's part and to deny that Darwin endorsed embryological recapitulation. That is simply wrong, Richards argues. Darwin was a recapitulationist and a progressivist – in particular ways and in ways that changed over time. Richards follows Darwin in suggesting that "when we regard every production of nature as one which has had a history . . . how far more interesting, I speak from experience, will the study of [the history of science] become!" (p. 84, this volume)

The second set of papers moves to this century and to the virtual explosion of laboratory and experimental research. The papers here explore the nature and use of evidence, considering such central questions as what counts as data, when data counts as evidence, and what role experimentation plays in revealing knowledge about living nature.

David Magnus revisits the naturalist-experimentalist distinction that Garland Allen and others have outlined on many occasions, and explains that what we have is really a case of competing epistemologies. Arguments by Hugo de Vries that changes in populations occur because of mutations, and that those mutations can even be large, conflicted with David Starr Jordan's conviction that it is isolation of parts of the population that leads to difference and change. This, and disputes like it, have been taken as indications of differences in theory about nature. And they have been interpreted as evidence for the existence of a naturalist approach to nature (with a more descriptive, qualitative, speculative, and evolutionary perspective) in contrast to an experimentalist approach (with a more quantitative perspective, more narrowly focused topics, and interest in reduction and micro-mechanisms). Magnus shows, by contrast, that what lies at the root is a disagreement about what counts as good science.

For Jordan, the naturalist, what matters in science, what he values, is consilience of a variety of lines of evidence, drawing on a diversity of methods. Breadth of theory and holism also are highly prized. Alternatively, de Vries and experimentalists value repeatability. It is not so much the experimentation per se as the possibility of repetition and hence definiteness that matters here. Experimentalists also look for parsimony, hence exhibiting a tendency to reduction, and the resulting rigor that they see in their approach and not in the naturalists' studies. This dispute is not, then, about whether to experiment or not. They all experiment. Rather, it is about the epistemological value of the experimentation. And this is not a tale about conflicting theories or debating individuals but rather a story about competing epistemologies, about what counts as knowledge and as good science.

What it means to be "right" in selected cases in developmental biology serves as the central question for *Jane Maienschein*. She argues that epistemological concerns actually drove the discussion in cases that have usually been taken as classic examples of theory conflict. Caspar Friedrich Wolff and Charles Bonnet in the eighteenth century, Wilhelm Roux and Hans Driesch at the end of the nineteenth, Camillo Golgi and Santiago Ramon y Cajal all argued about versions of preformation and epigenesis. Is the organismal form or the nerve already formed from the very first stages of development, or do they emerge later? In the case of Thomas Hunt Morgan, did he change his mind and move from an essentially epigenetic view to a genetic preformationism when he discovered the white-eyed male *Drosophila* fly, as the story has typically been told? No, these are not primarily conflicts over theory. Rather, they are more interesting stories about what should count as knowledge and about how to achieve "rightness" in biology; these are cases of competing epistemologies. Wolff and Bonnet argued about whether the senses can be trusted; Roux and Driesch about how much can be concluded from one counterexample; Golgi and Ramon y Cajal about which examples are decisive; and Morgan actually held to a consistent set of epistemic values, even while his theoretical interpretations changed significantly.

The electron microscope and PET (positron emission tomography) technologies raise further questions about how much observation can warrant belief. And which observations can we count? *William Bechtel* discusses these two examples and shows that ordinary perception is just the same. In the early stages of using new technologies, there is much concern about artifacts and about the reliability of both the techniques themselves and their interpretations. But the same concerns hold for ordinary perception, and we nonetheless have ways to deal with the concerns. We look to consiliences from other sources of evidence, and for consistency. Thus, we move beyond skepticism to a greater certainty of belief in our interpretations of what we are seeing. In general, then, Bechtel concludes that seeing really is believing – or that it becomes so as the field and technology develop.

The third section examines the nature and role of argument, including issues of object and styles. When there are competing epistemological frameworks, whether because of different styles of work or because of divergent ideologies, how do those differences play out in the science done as a result? The section considers issues of objectivity and other goals in science, and the way those have changed over time in response to a diversity of factors.

Larry Holmes reflects on the unpopularity of the logic of discovery in science among all but a few scholars such as Kenneth Schaffner. He wonders why, and asserts that historians need to work both to remain open to such considerations from philosophy and sociology of science and to retain their strong standards of historical study when looking at actual cases in detail. He focuses on Hans Krebs and his experimental work on the ornithine cycle of urea synthesis and on the citric acid, or Krebs, cycle. These represent "middle-range" theories that neither hold for all cases universally without exception nor are unique single instances. To what extent, Holmes asks, was the ornithine cycle discovered through logical processes? Was there a tight logic like the philosophers' propositional logic? No. But the process was logic and it was "intelligible to reason."

Holmes has followed through every step of Krebs's process, working from exquisitely detailed notebooks recording daily details of reasoning and experimentation. He records those details elsewhere, but here he explores the meaning of such work. What he sees is a process in which discovery and justification remain closely linked at all steps. There is, he explains, a tremendous interplay of thought and operations on a daily basis. Tiny steps of insight and creativity give way to the overall logic and reason. The process is effectively self-correcting, so that we see none of the great "breakthroughs" or revolutionary flashes or leaps that usually count as "discovery." Discovery we have, nonetheless. And it is that prolonged, reasonable process of daily experimental exploration that amounts to discovery.

Richard Lewontin provides a survey and substantial rethinking of issues from his book *The Genetic Basis of Evolutionary Change*. He concentrates on the difficulties standing in the way of knowledge in evolutionary genetics. To use a philosopher's word (one that Lewontin does not use), his subject is underdetermination. There is, he notes, a large number of basic biological mechanisms (forces), each of which enters in quantitatively quite different ways into the historical trajectories of different populations, or into the same trajectory at different times. This is due to differences in the organisms themselves, to external conditions, and to the stochastic nature of the operations of the mechanisms. The direct measurement of the forces is generally impossible, and such measurements of the populations as are available and that provide indirect information dramatically underdetermine the estimates of the quantitative values of those forces.

In view of this, it is perhaps unsurprising that the prospects for knowledge in evolutionary genetics are inversely proportional to the ambitions of the program designed to provide it. A maximal program aiming to provide the correct, universally applicable quantitative account of the forces of mutation, migration, selection, and breeding structures has little or no chance of success. More detail, but of less generality, is possible if we restrict ourselves to a model system. Still narrower is the demonstration that the limited evidence in a particular case is consistent with the occurrence of some theoretically possible process. In many cases the evidence is sufficiently weak that various parameters of a system cannot be independently measured. And where one is measurable, the others can be calculated only with the help of substantial assumptions sometimes amounting to the whole theoretical apparatus itself. Given such constraints, knowledge in evolutionary genetics will be a singular achievement and one that is likely to be highly restricted.

Marga Vicedo uses an example from early twentieth-century genetics to test claims by Ian Hacking (that we can know that certain theoretical entities exist because we can manipulate them in the lab) and by Nancy Cartwright (that we can know that certain theoretical entities enter into causal interactions in the lab and hence exist, even though we do not know the theories into which such entities figure). The example in question concerns William Castle's hooded rats. Castle bred tens of thousands of them at Harvard in order to determine whether observable differences depend on a malleable unit or factor of inheritance. He concluded that because he could breed the rats to exhibit a wide range of hoodedness, there were such malleable factors. E. M. East, by contrast, looked at the same data and concluded that no such malleable unit existed and that the phenomena were due instead to the simultaneous operation of many stable genetic factors. Vicedo shows that not only do the data of the lab get variably interpreted by scientists, they get variably interpreted and reported by historians of science as well. Vicedo concludes that the data do not speak for themselves. We cannot separate the question of whether the supposed entities enter into causal interactions from the question of whether the theories describing the interactions are acceptable. This is because on different theories different entities will be so involved. Similarly, we cannot evaluate the claim that we are indeed manipulating specific unobservable entities without choosing among theoretical accounts of such processes.

The core question of developmental biology, how the zygote becomes a multicelled organism, provides the focus for *Evelyn Fox Keller.* She argues that explanation is not self-evident in science but also functions locally and contingently as it meets the needs of the particular experimental system. Even funding interests may shape the formation of explanations. To make the point, she looks at genetic and epigenetic approaches to development. There are limits to the notion of genes as causes for development, notably the paradox that individual cells having the same genetic material (or information) develop differently. The discourse for gene action takes redundancy as a problem. Yet redundancy manifestly occurs, and epigenetic accounts take it as necessary for an acceptable explanation. Redundancy helps to produce the reliability and stability manifest in developmental systems. Through the examples, we see that the character of the quest for explanation, like the explanation itself, is both local and global, both contingent and contextual

Helen Longino examines five examples of apparent pluralism in biology. In each case, not only do different theories coexist, but the very questions at issue and the epistemic approaches addressing those questions command no consensus. Different data are taken as admissible or relevant; different epistemic or cognitive values are appealed to; and different arguments linking assumptions and practices to aims and goals are accepted. Rather than writing off biology as immature or unscientific, or writing off such epistemic concerns as sociologically irrelevant, Longino sketches a community-based picture of scientific justification which makes room for such pluralism. In this picture, which builds on her earlier work arguing for community-based objectivity, communities are constituted by a selection of substantive and methodological assumptions where the latter can be called the "local epistemology" of the community.

Finally, *Kenneth Schaffner's* "Afterword" ties together some of the emerging themes of the volume and relates them to recent literature in the history and philosophy of biology. Among the themes he highlights are controversies surrounding forms of empiricism and experimentation as well as issues of pluralism, discovery, and explanation in biology.

Taken together, these essays show the richness and diversity of studies in the biological sciences. They get at core questions in – and about – biology and show a range of philosophical concerns. They do not reveal a need for a philosophy of biology that is fundamentally unique, nor do they demand that study of life requires a special biological epistemology. Yet they do make clear that a close look at such diverse and focused cases within biology is important to allow us to reflect on the role of epistemology and other philosophical concerns. In addition, these studies, taken together, help to develop and deepen our understanding of how biology works and what counts as warranted knowledge and as legitimate approaches to the study of life.

Part One

The Nineteenth Century

Evolution and Its Contemporary Philosophy
of Science

Chapter 1

Darwin and the Philosophers

Epistemological Factors in the Development and Reception of the Theory of the *Origin of Species*

MICHAEL RUSE

Charles Darwin was a scientist, theoretical and empirical. Throughout his active career, he was forever speculating and throwing up new hypotheses. Equally, he was observing nature, experimenting, and, particularly toward the end of his life, coordinating the activities of friends and acquaintances toward his own needs and aims. It is my claim that, in this work, philosophical ideas played a major role. Most particularly, Darwin knew personally the major philosophers of science of his day. As was so often the case in other areas, Darwin responded fully to this exposure, making great use of these philosophers' ideas in his scientific activity. This fact influenced not only the work produced, but also its reception: specifically, the reception of Darwin's theory of evolution through natural selection, as expounded in his *Origin of Species* (1859). This helps to explain the well-known paradox that, although Darwin convinced his contemporaries of the truth of the fact of evolution, he was far less successful in convincing them of the truth of the particular mechanism he highlighted and endorsed (Ruse 1975b, 1979).

THE PHILOSOPHERS OF SCIENCE

In the fourth decade of the nineteenth century – that in which Victoria came to the throne – one sees in Britain a concerted effort to professionalize science, and to articulate a philosophy for the working scientist. The most notable mark of this activity was the founding in 1831 of the British Association for the Advancement of Science: a still-extant annual meeting which brings together people interested in science for a week of discussion and lectures. A meeting which, as

it happened, turned a tidy profit, thus enabling certain selected practitioners to receive funding for their activities (Morrell and Thackray
1981). Running in parallel with this almost commercial activity, one
sees attempts to upgrade the quality of debate at other scientific
institutions, most notably the Royal Society (Rudwick 1963). With
this, the tempo of improvement of British scientific education picked
up. It is true that it was not until the middle of the nineteenth century
that Oxford and Cambridge finally introduced science degrees; but
already those institutions were starting to offer sound courses in
various branches of the sciences. Moreover, there was the spur of the
newly founded colleges that were soon to make up the University of
London: all of these institutions included scientific education within
their curriculums.

Two men, Cambridge-educated, stand above all others as the the-
oreticians of science – those who were self-consciously articulating a
philosophy for the new professional science (Yeo 1993). The first was
John F. W. Herschel, who like his father the Astronomer Royal was
much concerned with the mapping of the heavens, but was interested also in many other branches of scientific activity. The second
was the Cambridge-based polymath William Whewell, later to become master of Trinity College, but at this time a fellow of the college
and well known as the author of standard texts on mathematics and
mathematical physics. Both Herschel and Whewell wrote works on
the philosophy of science: Herschel's *Preliminary Discourse on the
Study of Natural Philosophy,* appearing in 1830, and Whewell's mag-
isterial *Philosophy of the Inductive Sciences,* appearing in 1840, some
three years after his *History of the Inductive Sciences.* (In fact, there is
an unfinished manuscript on the philosophy of science by Whewell,
dating from the beginning of the decade. There is reason to think that
Whewell put this aside after Herschel's work appeared, and later
incorporated many of the ideas directly into *The Philosophy of the
Inductive Sciences.*)

There is significant overlap in the philosophies of Herschel and
Whewell. They both take physics in general, and Newtonian mechanics in particular, as the model of the best possible kind of science. For this reason, both men argue that the best science is what
today we would refer to as "hypothetico-deductive": they saw science as contained in axiom systems, with initial premises and with

lower-level theorems following from them (Hempel 1966; Popper 1959). Both saw the connection with the empirical world coming from testing the bottom-level theorems in Newtonian mechanics: particularly in the testing of the planetary laws of Kepler and the mechanical laws of Galileo against empirical experience. Thus, supposedly, the connection to the world (as it were) filtered back up through the system from the ground to the ceiling. Whewell (1829), in particular, was so taken with the axiomatic method of presenting science that he tried his hand at axiomatizing certain basic portions of political economy, with such success that today he is rather considered a forerunner of significant areas of the science.

However, there were fundamental differences between the two men, particularly over the nature of causation. The great Newton had somewhat cryptically argued that the best kind of science is causal. One would have expected him to say this, since his contribution to the scientific revolution was to provide a causal analysis of Copernicanism. But confusingly, Newton had left his followers in some doubt as to what would best constitute a causal explanation. He argued that one ought to strive not simply for causes but for what he called "true causes" (*verae causae*). It was here that Herschel and Whewell parted company, not so much over the need for causation per se, or even over the desirability of true causes. Rather, they differed as to the true nature of, and particularly the true way in which one identifies, *verae causae*.

Herschel stood very much within the British empiricist tradition, for all that his father had come over from Germany with the Hannovers. He argued that in order to know that one has a true cause, one must have either hands-on empirical experience of such a cause, or failing that, one must have direct empirical experience of something analogous. Thus, in explaining why we believe that there truly is a cause holding the moon in place as it spins around the earth, Herschel (1830) invited us to think of the analogy of a stone tied to the end of a piece of string, being swung around by hand. We know from physical experience that there is a force needed to pull this string in, as the stone spins around. Therefore, argued Herschel, we know what it is to say that we have a true causal understanding of the force acting on the moon to pull it in toward the earth, even as it spins around the earth.

Whewell (1840), who was influenced in his early years by Kant and then in the later years by Plato, argued much more strongly for what one might properly describe as a rationalist understanding of *verae causae*. He denied that one need necessarily have direct experience of a cause before one can label it a true cause. Indeed, if anything he rather distrusted direct experience. Rather, argued Whewell, the way in which we know that we have a true cause is that such a cause lies at the focus of a unifying explanation: it draws together our understanding in different areas of our experience. Perhaps, indeed, such a cause leads to new and unexpected predictions. Then and only then do we know that we have a true cause. Whewell's well-known phrase for the crucial unificatory experience is that of a "consilience of inductions." When everything points to one central cause, then truly one knows one has more than mere ad hoc supposition. One has a *vera causa* (Laudan 1981).

Neither Herschel nor Whewell was working in isolation, or merely at the level of theory. Indeed, there were genuine matters of scientific import motivating their discussions at this point. This was just the moment at which people were feeling that definitive proof had been offered establishing the veracity of the wave theory (udulatory theory) of light over the previously held Newtonian particle theory. The trouble was, of course, that no one could actually see the waves, or the particles for that matter. How then could one establish that, in undulating waves, one had a true cause? We find Herschel (1827) virtually bending over backward and pulling all sorts of contortions, as he supposed all kinds of visible and mechanical analogies to illustrate the action of the waves, particularly when they get into interference patterns: he proposed not only physical models involving waves, as for instance one finds in the bathtub, but all sorts of models involving sound waves coming off tuning forks dampened down with sealing wax and string and many other sorts of things. Thus, and only thus, did he think that it would be possible to show that light waves are *verae causae*. Whewell (1830), to the contrary, cut right through things, dismissing any need for physical analogues. As far as he was concerned, by supposing that light goes in waves, one can explain all sorts of varied phenomena in a consilient fashion. More than this, one can use the causal notion as a powerful predictive tool. This then suggests that waves are something really existing out there, rather than simply a figment of the theorist's imagination. For Whewell, the rationalist, the absurd analogical suppositions of Herschel, the empiricist, were quite unnecessary.

Another area where the *vera causa* question caused significant difference between Herschel and Whewell – a difference of more direct concern to us – concerned the nature of and causes behind the geological record. As is well known, 1830 was the year which saw the first volume of Charles Lyell's *Principles of Geology*. (The second volume appeared in 1832, and the third and final volume in 1833.) Lyell argued not only that the past can be explained in terms of causes akin to those of the present, but that one should expect to find the world in an ongoing steady state. To this end, he proposed his so-called "grand theory of climate," a theory which argues that the different temperatures of the earth and the variations over time are a function, not of a directional irreversible cooling phenomenon, but rather of the ever-changing distributions of land and sea. Such changes occur because the earth's surface is like a super water bed, in a constant state of rise and fall as such factors as erosion and deposition vary the forces pressing down on the earth.

Thus, against the directionalists' favorite piece of evidence, Lyell was able to argue that the tropical nature of the area around Paris, as revealed through the fossil record, is no true record of a cooling earth. It is simply a function of geographical distributions in the past that differed from those in the present. And in support of his case, Lyell argued that we have empirical evidence that distributions can make a difference: namely, the gulf stream. The distinctive flow of water up from the West Indies to the British Isles makes Britain a far more temperate place than its longitude would lead one to suppose or predict.

Herschel picked up on Lyell's theory of climate and – seeing just the kind of analogy he was seeking – used it enthusiastically in his little text on the philosophy of science as a perfect example of an empiricist *vera causa* in action. Whewell (1831, 1832) – who was, incidentally, responsible for naming the opposing positions, labeling Lyell a "uniformitarian" as opposed to the "catastrophists" who believed that the earth had a direction and was punctuated by upheavals – begged to differ. He argued that there is no need at all to search for empirical evidence today to justify one's positing of *verae causae* to explain the past. He argued, rather, that if the rocks suggest that there had been major upheavals, then this in itself is enough to justify one's assumptions that in the past there have been forces of a kind and nature unknown today. He concluded, therefore, that catastrophism is not only a preferable position to uniformitarianism, but that it and it alone truly satisfies the methodological demands of Newton with respect to *verae causae*. In the best rationalist tradition, catastrophism, which explains *to* the evidence, is to be preferred to uniformitarianism, which in empiricist fashion explains *from* the evidence. At this point, philosophical criteria are as important as the science itself.

CHARLES DARWIN AND THE *ORIGIN OF SPECIES*

Herschel and Whewell were not the only methodologists of science active in Britain at this point. There were others, including the Reverend Baden Powell (1855), Savillian Professor of Geometry at Oxford, and of course a little later the well-known John Stuart Mill, who published his *System of Logic* in 1843. (In fact, Mill was always far more a philosopher's philosopher than one directly concerned with the workings of empirical science, especially the physical sciences. Indeed, Mill by his own admission got most of his examples straight out of Whewell's writings.) But we have enough evidence now to ask our crucial question: Where and how do we fit Charles Darwin into this picture? What about Darwin and the philosophers of science?

Where we fit Darwin is an easy question to answer. Right in the middle! Darwin was a student at Cambridge from 1828 until 1831 (Browne 1995). While there he came into contact with the leading scientific men of the university, including Whewell, and we know from his later memories that this was a stimulating influence indeed. Then, right at the end of his university career, while putting in time before he could actually graduate, Darwin read with some care Herschel's just-published work on the philosophy of science, *The Preliminary Discourse on the Study of Natural Philosophy*. This was something very exciting and stimulating for the young man. Already Darwin was starting to think in terms of a scientific career, and Herschel's vision of science was something which moved him very much. Combined with this was the reading of Lyell's *Principles of*

Geology during Darwin's subsequent trip as de facto ship's naturalist on *HMS Beagle* (1831 through 1836). The first volume of the *Principles* was given as a present to the young Darwin by the captain, Robert Fitzroy; the later volumes were sent out to him from England. It is clear that what Darwin saw in Lyell was very much a reflection of what he had been taught to cherish by Herschel: causal science, empiricist science, the best kind of science. (Darwin visited Herschel in South Africa at the end of the voyage. The astronomer/philosopher was just then mapping the stars of the Southern Hemisphere.)

On returning to England in 1836, Darwin immersed himself in a heavy program of reading. Among other works, he read the three volumes of Whewell's *History of the Inductive Sciences* – quickly when it was first published at the beginning of 1837, and then in more detail, making annotations to the margins, later in the year. In addition, as soon as he returned to England, Darwin immersed himself in the scientific milieu of the day, joining the Royal Society and becoming active in the Geological Society, organizations in which Whewell played leading roles. (Whewell was president of the Geological Society from 1837 through 1839). We know that Darwin discussed methodological issues with him in some detail. I am not sure that Darwin ever read Whewell's *Philosophy of the Inductive Sciences;* but he read a very detailed review of the *Philosophy* and the *History* by John Herschel, published by *The Quarterly Review* in 1841 (Ruse 1979). This excited Darwin and interested him: hardly surprising, since within this review Herschel gave a detailed exposition of Whewell's thinking on the *vera causa* issue, bringing in the whole matter of the consilience of inductions. In all, there is incontrovertible evidence not only that Darwin knew Herschel and Whewell and their writings, but also that these were writings that interested him very much indeed.

We have answered the question of where we are to put Darwin into the picture. We turn now to the more difficult question of how we are to understand his position in the picture. Let us note some of the pertinent dates, to provide the proper background. It was early in 1837, some months after he returned from the *Beagle* voyage, that Darwin became an evolutionist (Sulloway 1982b). This occurred thanks particularly to the influence of John Gould, the bird systematist, who told Darwin that his specimens from the Galápagos Archi-

pelago (today we think it was probably the specimens of mocking-
bird) were unambiguously of different species (Sulloway 1982a).
Darwin at once realized that the only sound naturalistic explanation
for this phenomenon was evolution, and so he moved over the
divide to a commitment in common origins. He then worked fre-
netically for some eighteen months until at the end of September
1838. Having read the *Essay on a Principle of Population* by the Reverend Thomas Robert Malthus, he saw that a struggle for existence
would lead to a natural form of selection: thus, he had a causal
mechanism for the evolutionism to which he was now committed.

Darwin did not publish or announce anything at this time. But he
did continue to work very hard through the end of the 1830s and into
the beginning of the 1840s to embed his causal mechanism of selection within a full theory. In 1842 he wrote a thirty-five-page preliminary essay on the subject (now referred to as the "Sketch") and then
in 1844 he developed these ideas into a much longer 230-page essay
(now known as the "Essay") (Darwin and Wallace 1958). For some
ten years or so after this, Darwin was diverted into a massive study
of barnacles, living and fossilized. Then, in the mid 1850s, Darwin
returned to the species problem and started writing up a very large
book on natural selection. (This was unfinished and never published
by Darwin. It finally appeared in print in 1975 as *Natural Selection*.) In
1858, his hand was forced when a young naturalist and world trav-
eler, Alfred Russel Wallace, sent to Darwin (of all people!) a short
essay containing just those ideas on evolution and selection that
Darwin had himself discovered some twenty years previously. In
some fifteen months, Darwin rapidly wrote up his ideas, and the
Origin of Species was published at the end of 1859. This work was to
go through some six editions, the final and much augmented version
appearing in 1872.

I do not think that philosophy (from now on when I refer to
"philosophy" without qualification, I mean the philosophical ideas
of Herschel or Whewell or both) played a major role in Darwin's
becoming an evolutionist. What really counted here was the empirical evidence, particularly the already noted facts of biogeography. To
this should be added Darwin's background knowledge. He was,
after all, the grandson of Erasmus Darwin and had read his grandfather's major works (especially Darwin 1794). In addition, Darwin

had long known of the ideas of Lamarck ([1809] 1963). Also, a significant causal role should be given to religion or theology of some kind. By the time he became an evolutionist Darwin's religious faith had slipped from theism, the belief in an intervening God, to some kind of deism, the belief in a God who works through unbroken law. Evolution obviously is a manifestation and confirmation of this religious perspective (one which, I should note, was very much a Darwin family tradition). In addition to this deism, natural theological considerations played a major role. Darwin simply could not see how a sensible God would have put different organisms on the Galapagos Archipelago, virtually within sight of each other. On the South American mainland, where Darwin had just spent the last five years, one finds the same organism sometimes extended from the jungles of Brazil to the stony frigid deserts of Patagonia. Only a naturalistic explanation – and that meant evolution – could satisfy the demands that Darwin put on his God (Barrett et al. 1987).

Philosophy was not the spur. However, once he became an evolutionist, philosophy started to become significant. Darwin did not simply stop at becoming an evolutionist. Why did he feel the need to move on to a causal explanation, labors which took him eighteen months of hard thinking? Simply because his mentors had told him that the best kind of science is causal – and let us never forget that Darwin, as a fellow graduate of Cambridge University, would have taken very seriously the general assumption that the greatness of Newton lay in his providing a causal mechanism for Copernicanism. In addition, one should note that the kind of mechanism that Darwin postulated was very much akin to the kind of mechanism that Newton had postulated. Newton's key cause was that of gravitational attraction, some kind of force. Likewise, Darwin always thought of natural selection as being a biological equivalent of gravity, very much to be thought of as some sort of force, pushing and directing the organic world as Newtonian gravitation pushes and directs the inanimate world.

The next stage of Darwin's thinking likewise was indebted to his philosophical instruction. It was not enough simply to provide a cause: one had in some sense to embed it within a scientific theory. This is precisely what Darwin attempted over the next four to six years. It is true that Darwin hardly came up with anything like a rigorous axiomatic system, as is demanded by the hypothetico-deductive ideal. However, even here the axiomatic ideal was not irrelevant to Darwin's thinking. It is noteworthy how, right from the time of the "Sketch" through to the final version in the *Origin*, Darwin did not simply drop natural selection naked, as it were, onto the reader. Rather, he provided a little deductive argument to convince the reader of his mechanism's existence and significance. Always, Darwin started with initial Malthusian premises about the rapid (geometrical) rate at which organisms are reproduced and the limited (arithmetical) rate at which food and space can become available. (In fact, Darwin points out that space is absolutely limited.) This in turn leads, by inference, to the struggle for existence.

A struggle for existence inevitably follows from the high rate at which all organic beings tend to increase. Every being, which during its natural lifetime produces several eggs or seeds, must suffer destruction during some period of its life, and during some season or occasional year, otherwise on the principle of geometrical increase, its numbers would quickly become so inordinately great that no country could support the product. Hence, as more individuals are produced than can possibly survive, there must in every case be a struggle for existence, either one individual with another of the same species, or with the individuals of distinct species, or with the physical conditions of life. It is the doctrine of Malthus applied with manifold force to the whole animal and vegetable kingdoms; for in this case there can be no artificial increase of food, and no prudential restraint from marriage. (Darwin 1859: 63)

Then, from the struggle, Darwin moved on, adding another premise about the existence of variation in populations. Finally, deductively, he could infer the existence of natural selection.

Can the principle of selection, which we have seen is so potent in the hands of man, apply in nature? I think we shall see that it can act most effectually . . . Can it . . . be thought improbable, seeing that variations useful to man have undoubtedly occurred, that other variations useful in some way to each being in the great and complex battle of life, should sometimes occur in the course of thousands of generations? If such do occur, can we doubt (remembering that many more individuals are born than can possibly survive) that individuals having any advantage, however slight, over others, would have the best chance of surviving and of procreating their kind? On the other hand, we may feel sure that any variation in the least degree injurious would be rigidly destroyed. This preservation of favourable variations and rejection of injurious variations, I call Natural Selection. (Darwin 1859: 80–81)

Darwin, like almost all biologists of his era, was not very mathematical. So one should not expect to find, and indeed one does not find, any sustained attempt to extend into a rigorous axiomatic system the little set of inferences just discussed. There is nothing like the grand systems of the physical sciences. But, within the limits of his ability, one can certainly say that the way in which Darwin presents selection is much as one would expect to find from one who is taking seriously the methodological dictates of Herschel and Whewell.

But what about the question of true causes? Darwin believed that in natural selection he had found a cause: a causal force of a kind analogous to Newton's own central cause. In what sense, then, did Darwin feel that his force of selection could be offered to the world as a true cause? Herschel the empiricist and Whewell the rationalist differed on this. It is my claim that Darwin was sensitive to the ideas of both men and to their differences: he tried to cover his options, satisfying both the empiricist *vera causa* criteria and the rationalist *vera causa* criteria! In other words, Darwin tried to show natural selection to be a true cause as would be accepted by Herschel, and he also tried to show natural selection to be a true cause as would be accepted by Whewell.

The key to Darwin's attempt to satisfy the Herschelian empirical criterion is his use of the analogy of artificial selection. It is well known that, for Darwin, the parallel between the selection practices of animal and plant breeders and the natural processes one finds in the external world was extremely important. However, there has been considerable debate as to the exact nature of that importance. In his *Autobiography*, written toward the end of his life, Darwin claimed that it was artificial selection which gave him the clue as to the significance of natural selection, but that for some time he could not see how it would apply in nature. The breakthrough came only on reading Malthus and the description of the struggle for existence. Darwin then realized that a struggle could provide him the force or pressure necessary to fuel an ongoing natural selective process. However, some scholars doubt that this recollection was entirely accurate – perhaps artificial selection became important to Darwin only after he had discovered natural selection (Limoges 1970; Herbert 1971). Here, without getting bogged down in details, let me simply say that my own opinion is that Darwin's recollections were fairly close to the truth (Ruse 1975a). But whatever the truth about the question of discovery, when it came to the point of elaboration and presentation of the theory, no one can deny that Darwin used the artificial selection/natural selection analogy significantly.

In theory, it might have been that Darwin was simply using the analogy for pedagogical purposes: to prepare the reader for the unknown process of natural selection by first describing the known process of artificial selection. But in the *Origin* the analogy plays a much greater role than that (Ruse 1975a). It is intended in some sense to be a justification. Darwin tells the reader about artificial selection and assumes that the truth of this discussion will be accepted; then, on that basis, he tries to persuade the reader of the truth and force of natural selection. He does this, not just in a general way at the beginning of the *Origin*, but repeatedly throughout the work when he wants to convince the reader of some particular point of detail (Ruse 1979). Noteworthy, for instance, is his discussion of embryology: by invoking the case of the artificial world, where frequently the young of different adult forms are very similar because breeders have not selected for differences in the young, Darwin tries to persuade the reader that natural selection can be responsible for very different adult forms, despite the embryos' being very similar. Natural selection simply does not tear the embryos and young apart, as it does the adults. Life in the womb is the same for a dog as for a human.

One might think that there is nothing particularly noteworthy about the fact that Darwin uses the analogy in this way: it is not something which demands a particular comment. However, this is not quite true; or rather, this is the view from today when the analogy has become almost stale with familiarity. In Darwin's time, it was a daring move to make, almost to the point of recklessness, because *the* standard argument against evolution had always been based on the impossibility of changing one species into another. One might make bigger and better cows; one never makes cows into horses. Person after person had invoked the analogy in refutation of the evolutionary hypothesis. Indeed, this argument was such a commonplace in Darwin's day that the little essay Wallace sent to Darwin begins by invoking the analogy and arguing explicitly that it has no application to evolutionism! Wallace (1858) argued that, far from there being a significant analogy between artificial selection and natural selection, there are no true links at all, and that hence one should not take artificial selection to be a refutation of natural selection. Darwin turned this argument entirely on its head, arguing to the contrary that artificial selection proves or justifies our belief in the power of natural selection.

I do not say that Herschel's philosophy was the only influence at this point. Darwin almost uniquely among writers on the topic of evolution had a detailed knowledge of the practices of animal and plant breeders. The Darwin family had long kept pigeons, and as is well known Darwin himself became something of a pigeon fancier in midlife. In addition, his family – the branch headed by the man who was both his uncle and father-in-law, Josiah Wedgwood, Jr. – was deeply involved in agricultural improvements in England, improvements which involved the breeding of livestock and their fodder. Darwin therefore knew far better than most how strong artificial selection could be (Ruse 1979). But this in itself is hardly argument enough for the great use Darwin made of the domestic world. And here Herschel was surely significant: the skilled methodologist Darwin wanted some kind of analogy to his crucial causal force. Natural selection obviously is not something which we are going to see in action, at least not in any great way. No one saw natural selection bringing about the evolution of the dinosaurs, for example. However, with artificial selection Darwin had something which we do know about, of which we have had hands-on experience, and which he could therefore invoke to justify the belief in natural selection. He had in fact just what Herschelian *vera causa* criteria demanded.

Turning to the Whewellian side of the equation, a rationalist *vera causa* demands that one locate one's force at the center of a consilience of inductions. This is precisely what Darwin did! Having introduced natural selection in the fourth chapter of the *Origin*, Darwin devotes virtually the rest of the work to showing precisely how this selection can explain many different disparate parts of the biological world. Given natural selection, one can explain animal instinct; one can explain the facts of palaeontology; one can explain biogeographical distributions; one can explain morphology, embryology, systematics, and much more. Everything is brought together under the umbrella of selection; and conversely, as in a court of law when one appeals to circumstantial evidence to pin the blame on the defendant, so all of the subsidiary areas point to one true culprit: evolution through natural selection.

My argument, therefore, is that the structure of Darwin's theory of evolution – a structure to be found in the "Sketch" of 1842, the "Essay" of 1844, and in unchanged form in the *Origin of Species* of 1859, an argument that Darwin referred to as "one long argument" – is absolutely and entirely structured by the desideratum of building a theory with a *vera causa* at its heart. Moreover, my argument is that Darwin took seriously both the empiricist demands of John Herschel and the rationalist demands of William Whewell: we find in the structure of Darwin's theory the attempt to satisfy both sides to the *vera causa* equation.

Is there any proof of this claim, other than the fact that Darwin's work so clearly exhibits precisely what his acknowledged philosophical mentors were demanding? I admit that nowhere in his writings, published or unpublished, does Darwin come out and say explicitly that this is what he is trying to do. However, in addition to the evidence already before us, we do have unambiguous proof that Darwin thought systematically about his methodology, being quite self-conscious about wanting to argue for selection both analogically and through a consilience. Most famously, in response to one set of critical comments made after the *Origin* was published, Darwin openly defended what he had done in terms both of the analogy and of the sweeping, all-inclusive nature of his hypothesis.

In fact the belief in Natural Selection must at present be grounded entirely on general considerations. (1) On its being a *vera causa*, from the struggle for existence; and the certain geological fact that species do somehow change. (2) From the analogy of change under domestication by man's selection. (3) And chiefly from this view connecting under an intelligible point of view a host of facts. (Letter to G. Bentham, May 22, 1863; Darwin 1887: 3, 25)

Darwin went on thinking in these methodological terms. As we shall see shortly, Darwin was to be critiqued severely for his reliance on the artificial/natural selection analogy. At which point, he deliberately moved from one side of the *vera causa* wing to the other, arguing that even if one had no direct, hands-on evidence for selection, then it ought to be accepted on grounds of its consilient nature. Most interestingly and significantly, Darwin made reference to the wave theory of light, arguing that although we have no direct evidence of the wave theory, given its explanatory nature it is reasonable to accept the existence of such waves.

In scientific investigations it is permitted to invent any hypothesis, and if it explains various large and independent classes of facts it rises to the rank of a well-grounded theory. The undulations of the ether and even its existence are hypothetical, yet every one now admits the undulatory theory of light. The principle of natural selection may be looked at as a mere hypothesis, but rendered in some degree probable by what we positively know of the variability of organic beings in a state of nature, – by what we positively know of the struggle for existence, and the consequent almost inevitable preservation of favourable variations, – and from the analogical formation of domestic races. Now this hypothesis may be tested, – and this seems to me the only fair and legitimate manner of considering the whole question, – by trying whether it explains several large and independent classes of facts; such as the geological succession of organic beings, their distribution in past and present times, and their mutual affinities and homologies. If the principle of natural selection does explain these and other large bodies of facts, it ought to be received. (Darwin 1868: 1, 8–9)

I rest my case: Darwin had detailed knowledge of philosophical texts which argued that the ways in which he was to act were the ways in which one ought to act in order to produce good science. He took these writings seriously and was much impressed by them and their authors. Everything points to the conclusion that Darwin was influenced by these leading philosophers of science of his day.

THE RECEPTION OF DARWIN'S THEORY

How did the world receive the argument of the *Origin?* Speaking now primarily of Britain, but also of the intelligentsia in Europe, particularly Germany, and in the United States (at least in the northern states), we find that a few of the older thinkers simply refused to have much truck with evolutionism at all. Some, a very small minority – those who worked in areas where a causal mechanism was needed – embraced not just evolution but natural selection also. Noteworthy here was a man who had been a traveling companion of Wallace, Henry W. Bates (1863), who worked on problems of animal mimicry, particularly as found in butterflies. He became an enthusiastic Darwinian, in every sense of the word. The great majority became evolutionists very rapidly (Ellegard 1958); but although they agreed selection could have some power, they generally preferred other mechanisms: jumps or saltations, or some form of Lamarckism (the inheritance of acquired characteristics), or some kind of inner momentum (so-called orthogenesis), or some combination of these, or yet another kind of mechanism entirely (Bowler 1983, 1984).

Darwin sent complimentary copies of the *Origin*, with nice notes, to both Herschel and Whewell. Unfortunately, neither of them responded very positively. Whewell rejected outright the evolutionism of the *Origin*. Supposedly, he refused to allow a copy of the offensive work on the library shelves of Trinity College, of which he was now a master. (This may not be true, but it is a good story and quite plausible.) In the last edition (1863) of a work on natural theology which he had published exactly thirty years earlier (the *Bridgewater Treatise* on astronomy), Whewell took time in a new preface explicitly to reject evolutionism in general and selection in particular. By now, far sunk into a form of Christian Platonism, Whewell wanted no part of any naturalistic origins of the organic world, particularly not an organic world which includes us human beings. Herschel was a little more moderate, but not so that it made much difference. He referred disparagingly to Darwin's theory of evolution through selection as the "law of higgledy piggledy": a comment which, unsurprisingly, upset Darwin. In print, he was rather more complimentary, and from his unpublished correspondence it is clear that Herschel became an evolutionist of some kind. However, it was always a limited form of evolutionism, which postulated jumps or saltations from one major kind to another. It may even have been that Herschel thought that these saltations were in some way guided by the divine artificer, else he would not have been able to see how the designlike effects of the organic world could have come about. Herschel, like many others, was unconvinced by Darwin's claim that natural selection alone could explain adaptation and designlike function (Ruse 1979).

In a way, these reactions were no more than might have been expected. After all, these philosophers were in the "twilight" of their respective careers. What about others? In particular, what about the younger generation of scientific thinkers? What especially about those who gathered around Darwin and who venerated his achievements? Where did they stand on the theory of the *Origin* and to what extent was their stand influenced by philosophical considerations? Let us concentrate on the most central of these figures, the morphologist and paleontologist, teacher and self-avowed "Darwin's Bulldog," Thomas Henry Huxley. As is well known, Huxley's scientific position on evolution and its causes was somewhat paradoxical. Although he was an ardent supporter of Darwin at a personal level, and although he became the chief spokesperson for evolutionism as an idea, he was always somewhat hesitant about natural selection. He did not want to deny its power outright, but he certainly expressed a personal preference for other causes – specifically for saltations of some kind (Huxley 1893; L. Huxley 1900; Desmond 1994, 1997). And this was a preference from which he never deviated, even to the end of his life in 1895. Our question therefore is: Was this stand influenced by philosophical factors?

I claim that, although philosophy was certainly not the only factor in Huxley's attitude toward evolutionism, it was significant. At the general level of evolution itself, there were key philosophical factors which were important in Huxley's conversion. Intellectually, Huxley's philosophical allegiance was to a fairly broad form of empiricism. He even went so far as to write an enthusiastic book about David Hume (Huxley 1879). Along with this positive feeling, there was also opposition: opposition to the many idealistic currents of Huxley's day, represented most directly for Huxley by his arch morphological rival Richard Owen (1848). Against these idealists, Huxley argued that the only true stance for the working scientist is one which relates understanding directly to experience. In today's terms, this means that Huxley fell naturally into being what we would label a "naturalist": one who was determined, at all costs, to explain by reference to unbroken natural law. He wanted no truck with divine interventions or whatever (that is, with miracles). For Huxley, therefore, evolution was very much the epitome of his philosophy as applied to the organic world. To argue that organisms are as they are because of a long, slow, law-bound process was precisely what Huxley's empiricism demanded. (There are obvious links here to Darwin's having earlier accepted evolution because of his deism. However, although Huxley at times referred to himself as a scientific

Calvinist, by the time he accepted evolution he had drifted or matured into what he himself was to label "agnosticism": that is to say, a scepticism about any kind of religious entity or deity.

Combined with this empiricism, Huxley interlaced a strong dose of the most popular mid nineteenth-century ideological enthusiasm, namely that for progress: from simple to complex, from primitive to civilized, from monad to man. Huxley's closest friend during the 1850s was Herbert Spencer, from whom he learned much about evolutionism dressed up, as always with Spencer, in an overall metaphysic of upward progress: first to humans and then to English-speaking northern Europeans (Spencer 1857, 1868, 1904). For Huxley also, evolution took on this progressivist garb, and it is clear that he welcomed Darwin's message, less as a tool for the active scientist – what need has the morphologist of selection? – than as a kind of background metaphysical picture or secular religion against which he could perform his social acts. Which point shows that for Huxley there was an easy slide from the philosophical to the social: it was he, very much above all others, who was working hard in the second half of the nineteenth century to provide a place in British society for the secular professional scientist, a place where one would get a good education and then have the possibility of employment, publicly or privately financed. With some good reason, Huxley saw that his chief opposition was represented by conservative elements, epitomized by the philosophically idealistic Church of England. For Huxley, therefore, evolution was empiricist manna from heaven: it gave him and his fellows a kind of secular alternative – a progressivism-based alternative – to the Christian message of providential intervention and control. At this point, the philosophical and social became one.

What of natural selection and of Huxley's ambivalence bordering on rejection? There are many factors involved here, not the least of which was that just mentioned: as a morphologist, Huxley had no need in his scientific practice of a mechanism like selection. Morphology usually deals with organisms after they are dead; palaeontology, which was Huxley's later passion in life, deals with organisms only after they are dead! For Huxley, selection therefore had no attractions. Combined with this fact one must also add that Huxley, unlike Darwin, was untutored in and unimpressed by natural theology. The younger man never felt the need of a mechanism to explain the adaptive complexity of the organic world. Indeed, if anything, Huxley was indifferent to it: as a morphologist, what impressed him about the organic world, rather than its intricate adaptations, were the isomorphisms or homologies between the parts of very different organisms. In the language of the historians, Huxley was much more impressed by form than by function (Russell 1916). (Again, the fact that Huxley worked with dead organisms was significant. He never had to think about the ways in which organisms actually used their various characteristics. Indeed, for the morphologist, adaptation can be a distraction in working out the true conceptual links.)

And here again we swing around to the philosophical. Like everyone else, Huxley would have been looking for a genuine or true cause for the evolutionary mechanism. He was unimpressed by the rationalist claims of a philosopher like Whewell. In part, this was personal. Huxley saw Whewell as the enemy inasmuch as he represented the conservative establishment, quite apart from the fact that Whewell was Owen's idealistic mentor. (It was Huxley who told the tale about the *Origin* and the Trinity library.) It was also partly philosophical. The rationalist approach meant nothing to Huxley. Where would the morphologist find sweeping consiliences? What then of the empiricist side? Why was Huxley not impressed by Darwin's use of the artificial selection analogy? One has to feel that Huxley was not setting out to be impressed here. If he had wanted to make the analogy work, then he or some of his followers might have turned to experimentation to see if new species could be produced. (Huxley himself was never much of an experimenter, but he was good at motivating others to experiment. Particularly in the area of physiology, his students Michael Forster and H. N. Martin performed sterling deeds [Geison 1978].) But one senses that Huxley was not that bothered in the first place to prove selection effective. Hence, in the thirty-five years of his life after the *Origin* he was never motivated to work in that direction.

But having said this, it seems clear that the empiricism was a crucial factor in Huxley's reluctance. For Huxley, empiricism was more than just a methodology. It was very much bound up with the whole secular approach he was taking to the reformation of society, particularly in the area of education. Huxley's best-known saying came in his letter to Charles Kingsley on the occasion of the death of his (Huxley's) child, Noel. Huxley told Kingsley that one needed to sit down before the facts as a little child and let the truth speak to one. "Science seems to me to teach in the highest and strongest manner the great truth which is embodied in the Christian conception of entire surrender to the will of God. Sit down before fact as a little child, be prepared to give up every preconceived notion, follow humbly wherever and to whatever abysses nature leads, or you shall learn nothing." (Huxley 1900: 1, 219)

For Huxley, this was a directive not simply methodological, but in a sense sacramental: the hands-on experience with nature was something which Huxley saw as a kind of substitute for the Anglican Eucharist. Certainly, Huxley saw the empirical study of organisms as being something which, in the training of little Englishmen and women, would take the place of the sterile, stultifying (as he thought) study of the classics: dead books by dead authors. For Huxley, therefore, empiricism was a philosophy, but it was more than that. It was a motivating force for proper living in the deepest and most religious sense. Hence, inasmuch as selection at the artificial level failed to produce the effects that were being claimed for selection at the natural level – and what prospect do we ever have of turning a cow into a horse – this was more than just a matter for regret. It was a fundamental failing, and so Huxley turned his back so firmly on the whole process he could not even bring himself to make selection work.

We have therefore a remarkably well-formed package, one informed and shaped by philosophical ideas: acceptance of evolution as a matter of fact and as a background picture to one's overall thought, but rejection of natural selection as an effective mechanism of organic change. Given Huxley's great influence in education, particularly in Britain, but also in America (particularly through his student Martin [1876], who was one of the first biology professors at Johns Hopkins University), this was a vision which persisted for many years through several generations. Pertinently, the few people who challenged this picture, for instance the ardent Darwinian Raphael Weldon in England, realized that they had to take on Huxley on his own philosophical terms. Weldon, perhaps influenced by the positivism of his friend Pearson (1892), rightly saw that the empiricist would need to strengthen significantly the physical evidence of selection in action. He therefore engaged himself in selection experiments of considerable sophistication (Weldon 1898). It is true that he did not succeed in changing one species into another (nor, indeed, did he aim to do this), and perhaps this played a role in the less than universal enthusiasm with which Weldon's work was received. Indeed, although we today cherish Weldon's work (Gayon 1992; Depew and Weber 1994), it was not until the 1920s and 1930s that the Huxley picture (acceptance of evolution as fact and rejection of Darwinian selection as cause) was challenged seriously. By then, significantly, there were all sorts of other factors at work, not the least of which was a decline in enthusiasm for the extreme empiricism of the nineteenth century. In an age when physicists were proposing relativity at the macro level and quantum mechanics at the micro level, stringent demands that one's *verae causae* have direct reflections in experience no longer seemed compelling. But the implications of this – how new thinking about philosophy played a part in the revitalized Darwinism of the twentieth century – is a story for another place and another time.

REFERENCES

Barrett, P. H., P. J. Gautrey, S. Herbert, D. Kohn, and S. Smith, eds. 1987. "Darwin's Notebooks on the Transmutation of Species (Notebook B, P. 161)." In D. Kohn, ed., *Charles Darwin's Notebooks, 1836–1844*. Ithaca, N.Y.: Cornell University Press.

Bates, H. W. [1863] 1892. *The Naturalist on the River Amazon*. London: John Murray.

[1862] 1977. "Contributions to an Insect Fauna of the Amazon Valley." In P. H. Barrett, ed., *Collected Papers of Charles Darwin*, 87–92. Chicago: University of Chicago Press.

Bowler, P. J. 1983. *The Eclipse of Darwinism: Anti-Darwinism Evolution Theories in the Decades around 1900*. Baltimore: Johns Hopkins University Press.

1984. *Evolution: The History of the Idea*. Berkeley: University of California Press.

Browne, J. 1995. *Charles Darwin: Voyaging. Volume 1 of a Biography*. New York: Knopf.

Darwin, C. 1859. *On the Origin of Species*. London: John Murray.

1868. *The Variation of Animals and Plants under Domestication*. 2 vols. London: John Murray.

1958. Nora Barlow, ed. *The Autobiography of Charles Darwin, 1809–1882*. London: Collins.

Darwin, C. and A. R. Wallace. 1958. Foreword by G. de Beer. *Evolution by Natural Selection*. Cambridge: Cambridge University Press.

Darwin, E. [1794] 1996. "Zoonomia; or, The Laws of Organic Life." London: J. Johnson.

Depew D, and B. Weber. 1994. *Darwinism Evolving*. Cambridge, Mass.: MIT Press.

Desmond, A. 1994. *Huxley, the Devil's Disciple*. London: Michael Joseph.

1997. *Huxley, Evolution's High Priest*. London: Michael Joseph.

Ellegard, A. 1958. *Darwin and the General Reader*. Goteborg: Goteborgs Universitets Arsskrift.

Gayon, J. 1992. "Le Concept de Recapitulation a L'Epreuve de la Theorie Darwinienne de L'Evolution." In P. Mengal, ed., *Histoire du concept de recapitulation*. Paris: Masson.

1992. *Darwin et l'après-Darwin: Une histoire de l'hypothèse de sélection naturelle*. Paris: Kimé.

Geison, G. 1978. *Michael Foster and the Cambridge School of Physiology: The Scientific Enterprise in Late Victorian Society*. Princeton, N.J.: Princeton University Press.

Hempel, C. G. 1966. *Philosophy of Natural Science*. Englewood Cliffs, N.J.: Prentice-Hall.

Herbert, S. 1971. "Darwin, Malthus, and Selection." *Journal of the History of Biology* 4: 209–17.

Herschel, J. F. W. 1827. "Light." In E. Smedley et al., eds., *Encylopaedia Metropolitana*. London: J. Griffin.

1831. *Preliminary Discourse on the Study of Natural Philosophy*. London: Longman, Rees, Orme, Brown, and Green.

Huxley, L. 1900. *The Life and Letters of Thomas Henry Huxley*. (3 vols) London: Macmillan.

Huxley, T. H. 1879. *Hume*. London: Macmillan.

1893. *Collected Essays: Darwiniana*. London: Macmillan.

Lamarck, J. B. [1809] 1963. H. Elliot, trans. *Zoological Philosophy*. New York: Hafner.

Laudan, L. 1981. *Science and Hypothesis*. Dordrecht: D. Reidel.

Limoges, C. 1970. *La selection naturelle*. Paris: Presses Universitaires de France.

Lyell, C. 1830–1833. *Principles of Geology*. London: John Murray.

Malthus, T. R. [1826] 1914. *An Essay on the Principle of Population*. (sixth edition) London: Everyman.

Martin, H. N. [1876] 1967. "The Study and Teaching of Biology." (*Memoirs From the Biological Laboratory of the Johns Hopkins University* 3: 192–204) In W. Coleman, ed., *The Interpretation of Animal Form*, 181–91. New York: Johnson Reprint Co.

Mill, J. S. [1843] 1974. J. M. Robson, ed. *A System of Logic Ratiocinative and Inductive.* Toronto: University of Toronto Press.

Morrell, J. and A. Thackray. 1981. *Gentlemen of Science: Early Years of the British Association for the Advancement of Science.* Oxford: Oxford University Press.

Owen, R. 1848. *On the Archetype and Homologies of the Vertebrate Skeleton.* London: Voorst.

Pearson, K. 1892. *The Grammar of Science.* London: Walter Scott.

Popper, K. R. 1959. *The Logic of Scientific Discovery.* London: Hutchinson.

Powell, B. 1855. *Essays on the Spirit of the Inductive Philosophy.* London: Longman, Brown, Green, and Longmans.

Rudwick, M. J. S. 1963. "The Foundation of the Geological Society of London: Its Scheme for Cooperative Research and Its Struggle for Independence." *British Journal for the History of Science* 1: 325–55.

Ruse, M. 1975. "Darwin's Debt to Philosophy: An Examination of the Influence of the Philosophical Ideas of John F. W. Herschel and William Whewell on the Development of Charles Darwin's Theory of Evolution." *Studies in the History and Philosophy of Science* 6: 159–81.

1975a. "Charles Darwin and Artificial Selection." *Journal of the History of Ideas* 36: 339–50.

1975b. "Charles Darwin's Theory of Evolution: An Analysis." *Journal of the History of Biology* 8:219–41.

1979. *The Darwinian Revolution: Science Red in Tooth and Claw.* Chicago: University of Chicago Press.

Russell, E. S. 1916. *Form and Function: A Contribution to the History of Animal Morphology.* London: John Murray.

Spencer, H. 1868. *Essays: Scientific, Political, and Speculative.* London: Williams and Norgate.

[1857] 1868. "Progress: Its Law and Cause." *Westminster Review* 67: 244–67.

1904. *Autobiography.* London: Williams and Norgate.

Sulloway, F. J. 1982a. "Darwin and His Finches: The Evolution of a Legend." *Journal of the History of Biology* 15: 1–53.

1982b. "Darwin's Conversion: The Beagle Voyage and Its Aftermath." *Journal of the History of Biology* 15: 325–96.

Wallace, A. R. [1858] 1870. "On the Tendency of Varieties to Depart Indefinitely From the Original Type." *Journal of the Proceedings of the Linnean Society, Zoology* 3: 53–62.

Weldon, W. F. R. 1898. "Presidential Address to the Zoological Section of the British Association." *Transactions of the British Association,* 887–902.

Whewell, W. 1829. *On the Mathematical Exposition of Some Doctrines of Political Economy.* London: Gregg.

1831. "[Review of] Preliminary Discourse . . . by J. F. W. Herschel. . . ." *Quarterly Review* 45: 374–407.

LIBRARY
UNIVERSITY OF ST. FRANCIS
JOLIET, ILLINOIS

1832. "[Review of] Charles Lyell's Principles of Geology." *Quarterly Review* 47: 103–32.

1833. *Astronomy and General Physics.* (Bridgewater Treatise 3) London: Parker.

1837. *The History of the Inductive Sciences.* 3 vols. London: Parker.

1840. *The Philosophy of the Inductive Sciences.* 2 vols. London: Parker.

Yeo, R. 1993. *Defining Science: William Whewell, Natural Knowledge, and Public Debate in Early Victorian Britain.* Cambridge: Cambridge University Press.

Chapter 2

Knowing about Evolution

Darwin and His Theory of Natural Selection

JON HODGE

EVOLUTION AND COGNITION

On the topical overlap between evolution and cognition, there are various clusters of questions one might distinguish. Here are two such clusters. First, one might ask how knowledge of evolution can be achieved. Humans, after all, have been accumulating reliable records about observed changes in animal and plant species for only a few thousand years, while evolution has taken place over tens, hundreds, even thousands of millions of years. The human period, wherein alone direct experiential access to evolution has been possible, is a tiny moment suspended between a much vaster past and, it is hoped, an indefinitely prolonged future. Questions can, then, be raised about how anything can be learned about evolution given these severe experiential limitations. Second, one can ask what insights about knowing itself can be gained from considering our mental faculties as products of this evolutionary process. Perhaps less confidence, or perhaps more, should be put in those faculties if they are thought to be legacies from ancestral, animal adaptations, rather than supernatural gifts from an omniscient God who has made man in His own image.

Currently there are some customary divisions of labor regarding these two clusters of questions. Biologists are more likely to address the first, psychologists the second. Only philosophers are likely to feel responsible for taking them all on in an integrated way. Typically, those divisions of labor were less sharp in the nineteenth century than in our own time; and it is striking how the young Darwin, most explicitly in his notebooks from mid-1837 to mid-1839, ranges freely across both clusters. Not that this ranging is surprising, when one appreciates that he was, from mid-1838 on, keeping two distinct but allied series of notebooks. One series (B–E) may be called the zoo-nomical notebooks, as the first notebook carries the title *Zoonomia* (meaning the laws of life), while the other series (M–N) is devoted to "metaphysics" (meaning not, as it did originally, the theory of being, but rather what it had come to mean in the eighteenth century: namely, the theory of mind, including reason, will, consciousness, habits, the moral sense, the social instincts, and so on, in man and other animals).[1]

More precisely, Darwin's concern with both clusters of questions reached a peak in the summer and autumn months of 1838. But there is a difference. He had, as a geological disciple of Charles Lyell, been for some six years engaging questions about how we, although able to observe only the present, can have scientific theories that count as knowledge, not as mere conjecture, about the vast prehuman past recorded in the rocks. By contrast, his concern with the respective roles of innate instincts and of learned habits in the acquisition of knowledge, and so in the progress of science, was a more recent preoccupation. It had arisen as he had come to novel conclusions about how new habits could initiate adaptive changes in bodily structures.[2]

While acknowledging this difference, however, it is well worth focusing on those months as an exceptionally revealing phase in Darwin's intellectual development. For it is obviously instructive to examine how he had come to reach the views he held at that time; while it is no less instructive to reflect on how he went on from there, in the immediate and in the more distant future. It is instructive not least because it was very soon afterwards, in the winter of 1838–9, that he came to his theory of natural selection in its first full formulation, and also because the public, published Darwin of 1859 and thereafter turns out not to have changed his mind to any serious extent on either of those two clusters of questions about evolution and cognition, to use this anachronistic phrasing once more; nor, indeed, would he ever change his mind on these matters.

AGREEING WITH LYELL AND HERSCHEL

During the *Beagle* years, the most consequential commitment Darwin made was to become a zealous disciple, as he put it, of Lyell's views in geology. The way Lyell's views were structured in the three volumes of his *Principles of Geology* (1830–33) ensured that Darwin could not make that commitment without assimilating Lyell's own epistemological and methodological self-consciousness. This assimilation was aided and abetted by Darwin's admiration for John Herschel, Lyell's friend and Britain's leading astronomer and physicist. Darwin had read Herschel's *Preliminary Discourse on the Study of Natural Philosophy* (1830) before leaving on the voyage. He may well have been impressed then by Herschel's endorsement of Lyell's views. In the closing months of the voyage he met Herschel, now in South Africa for astronomical purposes; and they discussed geology as two knowing converts to Lyell's campaign.[3]

At the core of the consensus between Lyell and Herschel was the claim that geology could and should be, like celestial mechanics, a science that explained phenomena by reference to *verae causae* – that is, true, real, known, existing rather than hypothetical, conjectured, supposed causes. The consensus covered, moreover, the most contentious thesis of the *Principles:* namely, that this explanatory ideal could be satisfied only if very strong presumptions were made about the causes at work on the earth's surface throughout the temporal domain of geology, that is, throughout the vast aeons from the time when the oldest known fossil-bearing rocks were laid down through the present time and on into the future. The same causes, it was to be presumed, had been at work and with the same intensities and in the same overall circumstances, so that they have had and will continue to have the same sorts and sizes of effects. Only on this presumption could geology be a science of true causes, Lyell held. For a true cause, as Lyell and Herschel agreed, was one whose existence is evidenced independently of the facts it is invoked to explain. Here, they followed such explicators of Newtonian ideals of evidence and explanation as Thomas Reid in the previous century. The Cartesian vortices were adequate causes for planetary orbits. If they existed, then they would suffice to cause and so explain the orbits. But the trouble with the vortices was that the orbits themselves were the only facts that could be cited as evidence for the existence of these vortices. By contrast, Newton's gravitational force was evidenced by terrestrial phenomena, falling and swinging bodies, providing independent evidence distinct from the orbits in the heavens that it was invoked to explain. On the presumption of a uniformity of causes and laws, between the fallings and swingings down here and orbitings up there, the gravitational force could then be invoked as a true cause for those orbits.

It should be likewise in geology, Lyell and Herschel agreed. In this science, there should be a presumption of uniformity, not across space, from low to high, but across time, from the present that humans can observe to the vast prehuman, unobserved past that they were not present to observe. Geologists should refer the ancient changes recorded in the rocks to causes still active today and still adequate to produce, albeit often only over the long future ages, the same kinds and scales of effects. As their existence in the present can be confirmed independently of their action in the past, they are true causes, causes whose existence can be evidenced independently of their putative responsibility for the effects they are invoked to explain.[4]

Darwin's geological theorizing, from the middle of the voyage on, shows him to have knowingly embraced this *vera causa* evidential and explanatory ideal and to be knowingly conforming to it his theorizing on a whole array of phenomena: the extinction of large mammals in South America, the formation of coral islands, the distribution of erratic boulders, to name but a few. Nor would he ever abandon this ideal. Indeed, it would be tempting, at this point, simply to pass to his argumentation in the *Origin of Species* (1859) and to show how its structuring and so the composition of the book itself were deliberately conformed to the ideal. For, there, successive independent evidential cases are made for natural selection existing at present; for natural selection being adequate to produce over long ages new species from old; and for natural selection having been responsible for the production of those species now living and for those species that lived formerly and that have since become extinct.[5]

However, to pass directly from the private geological theorizing of the *Beagle* (1831–6) and post-*Beagle* London (1836–42) years, to the published biology of 1859, would be to miss out on the many intriguing biographical, historiographical, bibliographical, exegetical, and interpretative complexities that we are forced to confront if we concentrate, even if briefly, on the zoonomical and metaphysical inquiries as Darwin pursued and reflected upon them in those months of exceptional self-consciousness in the summer and autumn of 1838, just before the theory of natural selection was first conceived (not, as legend and Darwin's later memories have it, suddenly in September or October 1838, but over several weeks in very late November and December).

DISAGREEING WITH LYELL

One recurrent challenge Darwin faced, before and after the voyage, arose because he was knowingly disagreeing with Lyell about several successive theoretical issues while being unwilling to break with Lyell's most general views about the presumptions and ideals required for geology to hold its place as a high-ranking science matching celestial mechanics in its epistemological and methodological credentials. Most conspicuously, in 1835 he rejected Lyell's theory about the causes for species extinctions (competitive defeats occasioned by ecological disruptions) in favor of another theory (whereby, Darwin thought, species senesce and die of old age like individuals); this was a theory that Lyell had discussed but criticized as not having evidential support appropriate to a *vera causa;* and so Darwin deliberately sought such support, most particularly in accepted generalizations about the known senescence of plant graft successions.

In 1836 and even more so in 1837, however, he made two further breaks with Lyell that raised the challenge set by the *vera causa* ideal in a much more extensive way. In 1836 he was first tentatively inclined to suppose that new species were not as Lyell had held, independently created and too fixed in their characters to diverge into descendant species; for Darwin decided in favor of some species at least arising in the transmutation of others. In dismissing and rejecting such transmutations, Lyell had insisted that anyone who was considering it seriously should consider too the entire theoretical system of Jean Lamarck, transmutation's most notorious advocate. Darwin, in the spring of 1837, now completely convinced of transmutation following his return to England and decisive new expert judgments on his extant bird specimens and extinct fossil mammal remains, did precisely that. For he decided to take Lamarck's side against Lyell and to elaborate a comprehensive systemic structure of theorizing, incorporating the transmutation of species along with the spontaneous generation of infusorian monads, and including man in the progressive development from those earliest, simplest beginnings. It was a system deliberately designed to match the structure of Lamarck's system as represented in Lyell's exposition (rather than in Larmarck's own very different one). It is this new system that opens Darwin's Notebook B, under the title *Zoonomia* (borrowed from his own, transmutationist grandfather's most famous book title) in July 1837.

Darwin had then done, systemically, what Lyell had held any convinced transmutationist should do. However, Darwin clearly accepted that several of his system's components were far from *vera causa* kosher, so that the system as a whole was epistemologically and methodologically imperfect.

Darwin's awareness of its imperfection is registered in his very phrasing. As he sketches the system in the first two dozen pages of the notebook, he says sometimes that "we see" or that "we know" that something generally is the case; but he says of other generalizations that we "suppose" them to be so. Moreover, as he works to improve his system in the following days, he is plainly attempting as far as he can to replace supposing with seeing, conjecturing with knowing. For he works to find which suppositions (about monads, for instance) may have to be rejected because they have false consequences; which other suppositions have hitherto undiscerned explanatory advantages; and for which can be found evidential support that is independent of their explanatory advantages.

In and of themselves, these ways of improving any system of theory were commonplace and customary enough, even banal, one might have said. But that is the point. Darwin, while knowingly pursuing a subversive program of inquiry, was concerned to see how far he could make his evidential and explanatory argumentation conform to accepted standards.[6]

A PROSPECTIVE PROJECT AND "MY THEORY" (1837–8)

In subsequent weeks and months Darwin did not develop the system sketched in July 1837–did not, that is, develop it as such, as a whole system. Instead, he came to articulate two goals that emerged from improvements made to components of this system. One goal was the theory of species propagation, to use his own term. The successive, reiterated production of new species from old was likened to the growth by budding and branching of a tree. It was a tree with many twigs dying (representing species extinctions) while others are dividing (representing species multiplyings and divergings). The clusterings of twigs and the gaps left by dead ones represented the lesser differences among species of distinct genera, or the greater differences among species of distinct orders or classes. In this arboriform process, species formations were credited to geographical separations and so isolations, with adaptive divergence made possible by sexual reproduction with inbreeding in changed conditions. These, as Darwin was arguing in 1837, were the circumstances that allowed for one species to give rise to two or more distinct descendent species. His account of these circumstances, and his arguments for their existence and efficacy, constituted "my theory" of species formation.

As for the promissory project, it should be emphasized that this project was only ever projected. It was talked about as a prospect, but it was never pursued, let alone completed. Moreover, it would be quietly abandoned by Darwin at the very time, and not coincidentally, when the theory of natural selection became the new "my theory" in the winter of 1838–9.

The promissory project was to make use above all of certain biogeographical facts. For Darwin promised himself that he would assemble instances of geographical series of congeneric species, that is, several congeneric species spread out with no spatial gaps between them, but remaining distinct in their characters because not interbreeding and adapted to slightly different local conditions.

Thanks to the commitment to species transmutations, such geographical series could be taken to be the result of a temporal succession of changes in species. What the project would do was, therefore, to assemble such instances of species change, so that generalizations could be inferred; so that, in Darwin's words, the laws of change could be established. Finally, inquiry would proceed into the causes of change, the causes lawfully responsible for these lawful changes. Now the structure and strategy of this promissory project, as Darwin returns to it time and again over more than a year, manifestly combined Darwin's loyalty to Lyellian geology with his assimilation of a standard view of what a successful science, like Newtonian physics, most obviously comprises. Lyell, as always moving from the accessible present to the inaccessible past, had often used horizontal, geographical inquiries as a way to reach vertical, geological conclusions. His long chapters on the geographical distribution of animals and plant species had been explicitly designed on those lines. Most particularly, he had studied what limited species in their spatial extent at present, in order to decide what had limited their temporal durations over aeons of the past. Biogeography (including competitive, ecological relations) was to indicate the causes of species extinctions. Darwin's project was likewise designed to move from knowable geographical facts to conclusions, otherwise unknowable, about the laws and causes of change over time. It would, then, exemplify, extend, and indeed vindicate Lyell's reform of geology so as to satisfy the *vera causa* ideal. Equally, in constructing a three-layered pyramid with lawful causes at the top, laws of change in the middle, and individual facts about change at the bottom, it would match those successes in the physical sciences where, for instance, planetary orbital facts were referred to planetary orbital laws (such as Kepler's), which were in turn explained by reference to causal laws (such as Newton's laws of force).[7] However, this promissory project remained just that and so was developed no further by Darwin.

A THEORY AND ITS EVIDENTIAL AND EXPLANATORY CREDENTIALS

What Darwin does concentrate on in the spring of 1838 is developing his theory of adaptive species formations. He retains his invocations of the special powers of sexual (as opposed to asexual) modes of reproduction, of geographical isolation, and of inbreeding in changed conditions. But he adds now a new understanding of how reproductive isolation can emerge. He eventually decides that an analogy with races within domestic species indicates that conspecific varieties will in time acquire an instinctive aversion to interbreeding. The analogy depends not on a comparison so much as on a contrast between wild species and domestic races. For Darwin is convinced that domestication itself vitiates such instincts. He is prepared to argue, therefore, that, by contrast, when marked and prolonged varietal divergence of bodily structure takes place in nature, divergence as marked as the divergence shown by domestic dog races, it would be accompanied by an instinctive aversion to inter-breeding. This reproductive isolation would then allow further struc-tural divergence to proceed until interbreeding was no longer possi-ble even if the instincts were not preventing it. Associated with this whole line of argument were two others. First, reports on the results of crossing an old breed with a more recent one in dogs, say, sug-gested that characters become more strongly inherited and less likely to blend on crossing as time goes on: Yarrell's law, as Darwin dubs it after the name of his informant. Second, as with reproduction, so quite generally, Darwin thinks, changes in structure follow changes in instinct which in turn follow changes in habits: so, conversely, new habits, on becoming inherited and instinctive, lead through their new inherited effects on bodily structures to changed structures.[8]

These additions to his theory of species formation do not make it, in Darwin's judgment, completely secure evidentially. Especially, he notes that the most hypothetical part of the theory is its assumption that, in the wild, varieties of long standing will eventually cease to interbreed. He acknowledges that the best he can do on behalf of this assumption is to argue, analogically, from the evidence that, were their instincts not vitiated by domestication, then races of domesti-cated species would not readily breed together. Indeed, he thinks this line of proof for transmutation in the wild is one of his most original insights.[9]

During these summer months of 1838, Darwin often reflects on the epistemological and methodological strengths and weaknesses of his theorizing. The reflections invoke the standard staples of the day. It is, he notes, a virtue in a theory if it connects many otherwise disparate phenomena; it is good if a theory allows for successful predictions; and purely hypothetical conjectures should be replaced wherever possible. By themselves, these remarks would hardly con-stitute a comprehensive and consequential interpretation for Darwin to be giving his theoretical insights. But he does indeed integrate them into a coherent stance that he is explicitly resolving to act upon in composing a book he is already contemplating, a book on his theory of species propagations. The resolution is, moreover, twofold, in that his theory is seen as having two distinct virtues. First, there is evidence for the theory itself, independent of its explanatory virtue. Included in this evidence for the theory itself will be, for example, the support it has from those inquiries that compare and contrast species in the wild with races under domestication. Second, the explanatory virtue is to be displayed by showing how many facts of many kinds, from comparative anatomy, biogeography, palaeontology, and so on, can be given unifying, connecting explanations by referring them to the theory of species propagations as elaborated in the arbori-form representation of species multiplications, divergences, and extinctions.[10]

Here, then, in the summer of 1838, well before he has his theory of natural selection, Darwin is showing himself already committed to the twofold view that will eventually condition how the *Origin of Species* will be composed in 1859. For that book will be a rewriting of Darwin's unpublished "Essay" of 1844, itself an expansion of his unpublished "Sketch" of 1842; and the "Sketch" and "Essay" are explicitly divided, as the *Origin* is less explicitly, into two principal divisions: a first division setting out the evidence for natural selec-tion, that is, for its existence and its ability to produce and diversify species; and a second division arguing that many facts of many kinds, from biogeography, comparative anatomy, and so on, are best explained by that theory. These facts thus indicate that natural selec-tion was most probably responsible for producing the extant and the extinct species.[11]

The summer of 1838 is also a time when Darwin, who opened his metaphysics (M) notebook in July of that year, reflects more ex-plicitly than ever before or since on large issues of philosophy. He reaffirms his theism very conspicuously, but insists repeatedly that God acts in nature through regular laws and causes and not in excep-tional, miraculous interventions. Darwin declares himself a material-ist, insofar as that means that the workings of the mind are lawful, caused consequences of brain structures. What is more, he declares himself a determinist. There may be the appearance of material events happening by chance rather than in accord with universal laws, but the appearance is misleading, he holds; events ascribed to chance must be assumed to be due to hidden actions of regular causes. So too with free will; there is the illusion that necessitating causes are absent, but they should be assumed to be present, albeit hidden from our scrutiny.

Darwin's theism, materialism, and determinism were brought together in the conclusion that our possession of the very idea of the Deity is an inevitable consequence of the brain's organization and more remotely, therefore, of the laws responsible for that organization, laws instituted ultimately by God himself.[12] Accordingly, Darwin responded to William Whewell's *History of the Inductive Sciences* (3 vols, 1837) in one way that Whewell himself would never have welcomed. Having looked at it only briefly, it seems, the year before, on first acquiring it, Darwin now studied Whewell's book carefully in the late summer, through the autumn and almost up to Christmas 1838. Following Kant, Whewell had insisted that some very general principles, including those fundamental to science such as the principle of causation itself – the principle that every event has a cause–are presuppositions brought to the interpretation of experience, rather than conclusions drawn from experience. They are necessary and a priori rather than contingent and a posteriori. Darwin subsumed this view within his conviction that what is acquired as a habit can become an innate instinct by becoming hereditary. So principles that are necessary and a priori for us now were first encountered as general facts known a posteriori by our animal or even plant ancestors. In developing this account of human cognition, however, Darwin never considered that his own particular theories about plants or animals or coral islands were necessary or *a priori*. They continued to be empirically evidenced accounts of the causes producing particular kinds of effects in the living or in the physical world.[13]

Darwin registers various agreements and disagreements with Whewell's three volumes. For example, he likes Whewell's distinction between formal laws (like Kepler's) that introduce no causal powers and physical laws (like Newton's) that do so and can explain formal laws. This distinction fitted well with Darwin's understanding of how his promissory project would move beyond laws of change to causes of change. Whewell, however, made no dent in Darwin's old loyalties when he, Whewell, attacked Lyell's reforming of geology to meet the traditional Reidian *vera causa* ideal. Whewell argued that Lyell was mistaken in thinking that causes acting at present were in relevant ways better known than the causes whose effects the rocks recorded. Even more generally, Whewell argued that no successful sciences had succeeded by referring phenomena to known causes. If *verae causae* were taken, as they traditionally had been, to be known causes, then Lyell's understanding of how geology should secure its credentials as a science was, Whewell insisted, essentially misguided.[14]

Obviously, everything Darwin wrote about geology in subsequent years shows that he was entirely unmoved by Whewell's arguments and that he continued to side with Lyell and Herschel. Strikingly, an addendum to Darwin's *Journal of Researches*, written in November 1838, ends with an explicit reaffirmation of his conviction that to deviate from the Lyellian principles of geology would be to violate the very rules of inductive philosophy.[15]

In later publications, most notably his two-volume *Philosophy of the Inductive Sciences* of 1840, Whewell proposed an alternative to the traditional *vera causa* ideal. He called his new proposal the consilience of inductions. According to this proposal, one gives up any distinction between the evidence for a theory that is independent of its explanatory efficacy and evidence from that efficacy. For, Whewell argued, there can be no independent evidence in the sense usually intended. However, some dependent evidence can be strongly verifying of a theory. For, Whewell proposed, a theory is very unlikely to be false if it explains many facts of many kinds, including facts of a kind or kinds not contemplated when the theory was first conceived. Some writers fail to read Whewell correctly. Whewell knew perfectly well that explaining successfully many facts of many kinds had long been thought a virtue in a theory, as Herschel, for one, had emphasized. What was new in Whewell was the point about kinds of fact not contemplated when the theory was conceived.[16]

Consider next, then, two questions about Darwin. Did he learn about Whewellian consilience of inductions in 1838? And did he ever reject the old *vera causa* ideal and embrace that doctrine instead? As for the first, the answer is straightforward: No, he did not, because (despite the impression some historians have given) that doctrine is not taught in the *History* (1837) and so could not be learned from that source. As for the second question, yes, Darwin may well have read Whewell on consilience in the 1840s following publication of the

Philosophy. But did Darwin ever embrace and conform his evidential practices to that doctrine? One may fairly doubt it, for two reasons. First, most obviously in the *Origin*, Darwin is still distinguishing evidence for his theory independent of its explanatory efficacy from evidence that is not independent. Second, nowhere in that book or in any other, does Darwin ever tell the reader which kinds of facts were contemplated before and which after the theory was first conceived. So it is not simply that the *Origin*'s structure owes nothing to Whewell's distinctive teachings. In two ways, it was written in conformity with ideals Whewell sought to discredit and replace. The notion that the structure and strategy of the *Origin*'s argumentation is deeply indebted to Whewell's philosophy of science will doubtless continue to appeal to people, even though it simply cannot be reconciled with the great deal that is now known about the persistence of Darwin's adherence to Lyell's and Herschel's views from 1838 through to the writing and publishing of that book.[17]

A NEW THEORY MEETS AN OLD IDEAL

When Darwin came to his theory of natural selection, over a period of several weeks from late November through to early 1839, much was retained that was in the earlier theory of species formation. A decisive shift occurred, however, in the comparing and contrasting of wild species and domesticated races. Contrary to previous contrasts, both were now interpreted as adaptations. For both were explained as adapted by selective breeding. Further, both man's and nature's selection were thought to be able to produce adaptations even when working with chance or accidental variations. The two selections were thought to differ in degree: nature's was far more precise, prolonged, and comprehensive (affecting many characters, that is, rather than a few). So Darwin could say that species formed by nature and races made by man were produced by the same means, albeit very different in degree. A proportionality could then be argued for: nature's selection was to man's as species are to races. The causes are proportioned as the effects are. The powers and effects of man's selection being known, those of natural selection are to be inferred in accord with this proportionality. As for the existence of selection in the wild, a process of selective breeding is entailed by heredity, variation – especially in the changing conditions geology testifies to – and the superfecundity dramatized by Malthus.[18]

Thanks to its new selection analogy the new "my theory" had, for both adaptation and species formation, better *vera causa* credentials than had the earlier theory, and Darwin's notes show him cherishing it on that account.[19]

Where he is less explicit, indeed silent, is on how the new theory relates to the old promissory laws and causes of change project. It may well be that the old project drops away as the new theory comes into prominence, because the new theory is seen to make the old project redundant, even misconstrued. The old project would have used present geography as a foundation for knowledge of the course and causes of past change. The new theory finds another way to move from the present to the past, for it moves, by analogy, by proportion, from what man does with selection in the short run to what nature does in a vastly longer run. The gap between the observationally known present and the unknown past is crossed in a new way. But there is a peculiar feature of natural selection as a cause. It is lawful in that it arises in a regular fashion from the natural powers of heredity, variation, and superfecundity. It has, however, no law of its own; for there is no single, universal law that is to natural selection as the Newtonian inverse square law is to gravitational force. In that sense, natural selection is a cause that has no one law of its action. So the old promissory project's aim to find laws of change is hardly satisfied; it is rather circumvented, insofar as the new theory as to the cause of species changes includes no law for species changes. This lack of a law was, Darwin's later critics would urge, a difficulty. With gravitational theory, the law allows one to deduce what effects the cause will produce in specified circumstances, so that one can then decide whether these predictions are born out in fact. Natural selection does not seem susceptible to confirmation in this way.[20]

Any account of how Darwin's epistemological and methodological commitments were acted upon in the 1860s, especially in his public defenses of his theorizing, would properly require a paper in itself. But it may be worth making some brief comments here. First, any careful exegesis of Darwin's scientific argumentation will show, in my view, that he only ever had one ideal in mind, the one he saw Lyell and Herschel jointly upholding. What differs, then, is how fully, in any case, Darwin thought that ideal could be met. He certainly thought the *Origin*'s exposition met it fairly well. But he was emphatic that his evidential case for pangenesis met it far less successfully, most obviously because while heredity, variation, and superfecundity were known, observable powers of plants and animals, the existence of the pangenes was not at all susceptible to anything like direct confirmation. The presentation of pangenesis accordingly does not match the *Origin* very closely at all.

Second, faced with objections to the arguments of the *Origin*, Darwin often appears in print and in private correspondence ready to retreat and even sometimes to abandon the claim that he is meeting the old *vera causa* ideal. For he seems to set aside that ideal as inappropriately demanding and to argue that some very reputable theories in physics, such as the wave theory of light, do not confirm to that standard but are nevertheless widely agreed to be little, perhaps none, the worse for not doing so. Natural selection should be then judged, he appears to suggest, by those other standards that have been invoked in making the case for, say, that optical theory. In making these moves, Darwin puts more emphasis on the virtue of explanatory unification and less on the virtue of evidence independent of that efficacy; and this emphasis takes him nearer to Whewell, further away from Lyell and Herschel.

That Darwin makes some such moves is manifest from the relevant familiar texts. But it is worth reading those texts with two questions in mind. First, how wholeheartedly does Darwin repudiate the appropriateness of the old *vera causa* ideal for his theory of natural selection? Is it not true that he continues to think that a defender of natural selection should hope that people will eventually obtain the evidence that would show natural selection to be a *vera causa*? For Darwin never suggests that he and his allies should not try to make stronger the evidential cases for the existence and adequacy of natural selection, the cases set out in the early chapters of the *Origin*, the chapters preceding those later ones devoted to the theory's explanatory virtue. Now, if Darwin had really come round to Whewell's own rejection of the old *vera causa* ideal in favor of the consilience of inductions, if Darwin had really undergone such a fundamental shift in his understanding of what makes the best scientific theorizing epistemologically and methodologically superior to the rest, then there would be far more dramatic shifts than are to be seen in his beliefs about where the way forward lay in making his theory a permanent, accepted contribution to science.

Finally, when one reviews the full sweep of Darwin's theorizing, whether in geology, biology, or psychology, there is one general conclusion that is surely difficult to resist. In his theorizing he was often, although not always, not merely markedly innovative but radical and subversive, and knowingly and deliberately so. But in his own eyes this theorizing did not require him to be correspondingly innovative about the epistemology and methodology of science. On the contrary, he seems anxious that his new theories be developed and presented within the familiar, reputable constraints satisfied by what his contemporaries count as the best science of the day. A moldbuster yes; but, in his own eyes not a metamoldbuster. Obviously, one could argue that Darwin was mistaken and misguided on this issue of how his scientific theories related to the contemporary philosophical theories about scientific theories. Perhaps, in effect if not in intention, he was truly, or at least has turned out to have been truly, a buster of molds at both levels. That is an exciting claim, but our excitement should not lead us to read back, into Darwin's own life and work, too many of the radical and innovative philosophical lessons that have been drawn from evolutionary biology in the century since Darwin. There is no danger that anyone will ever show Darwin to be a completely dull fellow. The notebooks, if not the published texts, show what a lively, imaginative, even wild and weird mind the man had. But there is no need to represent him as upsetting everything at every level. Nor, conversely, is there any need to make Darwin a canonical authority imposing limitations on our own reflections about evolution and cognition. The uses of Darwin for the purposes of philosophical inspiration can surely coexist peacefully, even fruitfully, with the historian's Darwin.[21]

NOTES

1. These notebooks are now available in Barrett et al. (1987). References here will be by notebook and manuscript page number, as in that edition. For recent accounts of these years in Darwin's life and work see the two recent biographies, which include extensive reference to the specialist

literature: Desmond and Moore (1992) and Browne (1995). The state of
the art in Darwinian scholarship fifteen years ago is accessible in Kohn
(1985). A main location for Darwinian studies since then is *The Journal of
the History of Biology.*

2. Hodge (1982), Hodge and Kohn (1985), Richards (1987), Curtis (1987).
3. Herschel (1830), Ruse (1975).
4. See Hodge (1977, 1982), Ruse (1975), and Laudan (1987) for fuller accounts of Lyell's views in relation to Herschel's.
5. Hodge (1977, 1982, 1989)
6. For this whole section, see Hodge (1982) and Darwin's Notebook B 1-24.
7. For this section, see Notebooks B and C. For the last general exposition of the promissory project, see Notebook E 51–55.
8. See Notebook C and Hodge and Kohn (1985).
9. See Notebooks C 30, C 176-177, and D 69.
10. Notebooks B 104, C 62, C 76-77, D 69, D 71, D 117.
11. Hodge (1977, 1989).
12. See Notebook M, Richards (1987), and Manier (1978). See also Notebook C 166.
13. The indexes to Barrett et al. (1987) allow one to see when Darwin was reading Whewell. Curtis (1987) has a much fuller discussion than can be given here of Darwin's response to Whewell's epistemology for science. See also Curtis (1986).
14. Darwin's annotations of Whewell's *History* are given in Di Gregorio (1990) 866–8. Whewell's rejection of Lyell's reforms for geology is in the final chapter of Whewell's third volume. On Whewell and Lyell, see also Ruse (1975), Laudan (1987), and Hodge (1991).
15. Darwin (1839) 615–625. In this addendum, Darwin writes in a highly self-conscious and self-serving way about the epistemological superiority of his explanation for erratic boulders (rafts of floating ice) over a rival explanation (diluvial debacles) proposed by Louis Agassiz. Darwin (625) insists that, in his explanation, "only *verae causae* are introduced," and, what is more, "reasons can be assigned, for the belief that these causes have been in action" in the districts in question. By contrast, to resort to a deluge hypothesis before it is "absolutely forced on us," is, he claims, "to violate, as it appears to me, every rule of inductive philosophy."
16. See Laudan (1982) and Butts (1973).
17. Indeed, the notion is defended in this very volume, in Michael Ruse's enlightening essay. I can only insist that readers of our two pieces, in making up their minds about Darwin's debts to Whewell, should not overlook our agreements on other issues.
18. See Notebooks D and E and Hodge and Kohn (1985).
19. See, for example, Notebook E 71, 118, and Darwin's notes on a theologi-

cal book by John Macculloch printed in Barrett et al. (1987) 632–641, especially MS pp. 53 and 57.

20. See Kitcher (1985).

21. These last three paragraphs take us into two issues that cannot be done justice here. The first issue is whether Darwin's later epistemological and methodological reflections, in print and in private correspondence after the *Origin* was published, show him to have moved away from the views of Lyell and Herschel and toward those of Whewell. The second issue concerns how far J. S. Mill, Whewell, and other philosophers of science saw Darwin's theory of natural selection as satisfying their notions of what good scientific theorizing and reasoning looks like. The second issue I will leave aside here. It is best explored by starting with David Hull's essay in this volume, and then going to Hull (1995) and Hull (1975).

The first issue involves the question of when Darwin read and reflected on Whewell's views, especially about the consilience of inductions. Here Thagard (1977) makes a natural point of departure. But a separate question has to be raised about the timing of any shift that Darwin is thought to have made. Consider, for example, the well-known passage, quoted in this volume by Michael Ruse, from the introduction to Darwin's treatise *Variation of Animals and Plants under Domestication,* first published in 1868. That passage may indeed seem to invoke Whewell's views. However, in 1860 Darwin had already written in almost the same words, and to almost exactly the same effect, in a letter to S. P. Woodward. And yet three years on, in 1863, writing to George Bentham, he takes a line which is much closer to the old consensus between Lyell and Herschel. (For the letter to Woodward, see Burkhardt et al. 1993, 123; for the letter to Bentham, see Darwin 1888, vol. 3, 24–5 or the quotations from it given in Hodge 1977 and Hodge 1989). When these and other similar texts (for one other, see Hodge 1989) are all taken into account, it seems that the apparent inconsistencies in them can be largely, if not fully, resolved by distinguishing, on the one hand, Darwin's own view of the credentials he thought his theory ought to have and could have, and, on the other, the view that he thought readers would do well to take of the theory's credentials.

Thus, Darwin really did think that he had shown, through the analogy with man's selection, that natural selection could cause intersterile species, even if man's selection had not ever produced intersterile races of a domestic species. But many critics insisted that this line of evidence failed to establish natural selection as an adequate cause for intersterile species formation; and so, to that extent, the hypothesis that extant and extinct species had been produced by selection was just that, a mere hypothesis. To anyone who took this view, Darwin would insist that, since no one had shown that selection could not produce new species from old, it was at least only fair to see how much this hypothesis could explain of geographical, embryological facts, and especially to see whether it could offer a more satisfactory explanation than the alternative hypothesis of separate creations of fixed species. So, in short, Darwin himself never really thought that he had only succeeded in formulating and defending a hypothesis that had no better *vera causa* credentials than the wave theory of light, which he knew was widely held to have almost none. But Darwin was happy to invite other people to see it that way, provided they played fair and compared its explanatory virtues with the special creationist alternative. In making that invitation, he knew enough to know that he was inviting other people to look at the theory more as Whewell looked at the wave theory of light, and less as Reid, Lyell, and Herschel had viewed the Newtonian theory of gravitational attraction. However, he never accepted his own invitation to look at the theory that way. He remained much too firmly convinced that his theory of natural selection satisfied the standards he had first embraced before he had even formulated the theory. (For a recent discussion of the structure and strategy of the argument of the *Origin* with references to earlier accounts, see Hodge 1992). Finally, let me emphasize what is implicit in this chapter as a whole: namely that Darwin was much more radical and innovative regarding the second cluster of questions (as distinguished in my first section) than regarding the first cluster. Curtis (1986 and 1987) sees him as equally so regarding both – an exciting but unsustainable proposal, I submit.

REFERENCES

Barnett, P. H., P. J. Gautrey, S. Herbert, D. Kohn, and S. Smith, eds. 1987. *Charles Darwin's Notebooks, 1836–1844.* Cambridge: Cambridge University Press.

Browne, J. 1995. *Charles Darwin,* volume 1. New York: Knopf.

Burkhardt, F., D. M. Porter, J. Browne, and M. Richmond, eds. 1993. *The Correspondence of Charles Darwin, volume 8* Cambridge: Cambridge University Press.

Butts, R. 1973. "Reply to David Wilson: Was Whewell Interested in True Causes?" *Philosophy of Science* 40: 125–128.

Curtis, R. C. 1986. "Are Methodologies Theories of Rationality?" *The British Journal for the Philosophy of Science* 37: 135–161.

  1987. "Darwin as an Epistemologist." *Annals of Science* 44: 379–408.

Darwin, C. 1839. *Journal of Researches into the Geology and Natural History of the Various Countries Visited by H.M.S. Beagle.* London: H. Colburn. Reprinted in facsimile (1952), New York: Hafner.

  1859. *On the Origin of Species* London: John Murray. Reprinted in facsimile (1975), Cambridge, Mass: Harvard University Press.

1888. *The Life and Letters of Charles Darwin*. 3 vols. Edited by F. Darwin. London: John Murray.

Desmond, A. and J. Moore. 1992. *Darwin*. London: Michael Joseph.

Di Gregorio, M. A. 1990. *Charles Darwin's Marginalia*, volume 1. New York and London: Garland Publishing.

Herschel, J. F. W. 1830. *A Preliminary Discourse on the Study of Natural Philosophy*. London: Longman. Reprinted in facsimile (1987), Chicago: University of Chicago Press.

Hodge, M. J. S. 1977. "The Structure and Strategy of Darwin's 'Long Argument.'" *The British Journal for the History of Science* 10: 237–246.

1982. "Darwin and the Laws of the Animate Part of the Terrestrial System (1835–1837): On the Lyellian Origins of His Explanatory Program. *Studies in the History of Biology* 6: 1–106.

1989. "Darwin's Theory and Darwin's Argument." In M. Ruse, ed., *What the Philosophy of Biology Is: Essays Dedicated to David Hull*, 163–182. Dordrecht: Kluwer Academic Publishers.

1991. "The History of the Earth, Life and Man: Whewell and Palaetiological Science." In M. Fisch and S. Schaffer, eds., *William Whewell: A Composite Portrait*, 255–289. Oxford: Clarendon Press.

1992. "Discussion: Darwin's Argument in the *Origin*." *Philosophy of Science* 59: 461–464.

Hodge, M. J. S. and Kohn, D. 1985. "The Immediate Origins of Natural Selection." In D. Kohn, ed., *The Darwinian Heritage*, 185–206. Princeton: Princeton University Press.

Hull, D. 1973. *Darwin and His Critics*. Cambridge, Mass.: Harvard University Press.

1995. "Die Rezeption von Darwins Evolutionstheorie bei britischen Wissenschaftsphilosophen des 19. Jahrhunderts." In E- M. Engels, ed., *Die Rezeption von Evolutionstheorien im 19. Jahrhundert*, 67–105. Frankfurt: Suhrkamp.

Kitcher, P. 1985. "Darwin's Achievement." In N. Rescher, ed., *Reason and Rationality in Science*, 127–189. Washington D.C.: University Press of America.

Kohn, D. 1985. *The Darwinian Heritage*. Princeton: Princeton University Press.

Laudan, L. 1981. *Science and Hypothesis: Historical Essays on Scientific Methodology*. Dordrecht: D. Reidel.

Laudan, R. 1987. *From Mineralogy to Geology: The Foundations of Science, 1650–1830*. Chicago: University of Chicago Press.

Lyell, C. 1830–33. *The Principles of Geology*. 3 vols. London: John Murray. Reprinted in facsimile (1991), Chicago: University of Chicago Press.

Manier, E. 1978. *The Young Darwin and His Cultural Circle*. Dordrecht: D. Reidel.

Richards, R. J. *Darwin and the Emergence of Evolutionary Theories of Mind and Behavior*. Chicago: University of Chicago Press.

Ruse, M. 1975. "Darwin's Debt to Philosophy: An Examination of the Influence of the Philosophical Ideas of John F. W. Herschel and William Whewell on the Development of Charles Darwin's Theory of Evolution." *Studies in the History and Philosophy of Science* 6: 159–181.

Thagard, P. 1977. "Discussion: Darwin and Whewell." *Studies in History and Philosophy of Science* 8: 353–356.

Whewell, W. 1837. *History of Inductive Sciences.* 3 vols. London: J. W. Parker.

Chapter 3

Why Did Darwin Fail? The Role of John Stuart Mill

DAVID L. HULL

The purpose of this paper is to explore the interplay between science and the philosophy of science. What effects do scientific ideas have on the philosophy of science, and conversely, what effects do philosophical views about science have on science? More narrowly, I investigate the effects that John Stuart Mill's philosophy of science had on the reception of Darwin's theory of evolution, and vice versa. Darwin began his career right when philosophy of science as a separate discipline was getting started in England. Later, when he published his *Origin of Species* (1859), his theory was evaluated on the basis of these contemporary philosophies of science (for early discussions of the relationship between science and philosophy of science, see Hull 1972, Ruse 1975, Laudan 1977, and more recently Engels 1995).

Students of science frequently ask why Darwin succeeded in establishing his theory of evolution, when his predecessors had failed so miserably. However, if by "succeed" one means that large numbers of Darwin's contemporaries came to accept the fundamentals of his theory of evolution, then Darwin did not succeed; he failed. I have yet to find a single major figure at the time who accepted *all* of Darwin's fundamental principles, for example, that variations occur in all directions (usually referred to as "chance variations"), that evolution is gradual, and that natural selection is the major directive force in evolution (Hull 1985). Only when one restricts oneself to the minimal claim that species in some sense change through time did Darwin make much of an inroad on the views of his day. According to one sample, 75 percent of Darwin's fellow scientists accepted the minimal claim that species "evolve" by 1869 (Hull, Tessner, and Diamond 1978).

Thus, Darwin succeeded to a large extent in convincing his contemporaries that species evolve but failed in persuading them that
species evolve in the way that he thought they did. That Darwin
provided an entirely naturalistic mechanism for the evolution of
species contributed to the acceptance of the idea of species evolving
even though, paradoxically, few of his contemporaries accepted his
entirely naturalistic mechanism. Instead, they formulated versions of
evolutionary theory that did not clash so markedly with their views
of nature and of science. The writings of John Stuart Mill played an
important role in these revisions.

NINETEENTH-CENTURY PHILOSOPHY OF SCIENCE

The chief founders of philosophy of science in Victorian England
were John Herschel (1792–1871), Charles Lyell (1797–1875), William
Whewell (1794–1866), and John Stuart Mill (1806–1873). John
Herschel's *Preliminary Discourse on the Study of Natural Philosophy*
appeared in 1830. Soon thereafter Charles Lyell published a general
philosophy of science in his *Principles of Geology* (1830–33), a philosophy, needless to say, that was consonant with his principles of geology. William Whewell published his three-volume *History of the Inductive Sciences* in 1837, followed in 1840 by two volumes of his
Philosophy of the Inductive Sciences, Founded upon Their History. Finally,
John Stuart Mill published his *A System of Logic* in 1843.

Throughout this period, these authors reviewed each others'
work, trying to clarify where they differed and, when they differed,
why the other person was mistaken.[1] All of these authors tried to
sound as empirical as possible. The good word at the time was
"induction," and the patron saints of induction were supposedly
Bacon and Newton. All of the early British philosophers of science
tried to enlist these great figures as their predecessors, even when
their views departed sharply from those of these authorities. Public
relation ploys to one side, the only early philosopher of science who
actually held views that were as inductive as he tried to make them
sound was Mill. If Mill had not existed, historians of philosophy
would have had to invent him.

As Ruse (1975, 1976a) has remarked, the inductive methods proposed by Herschel and Whewell were roughly equivalent to what

we now term "hypothetico-deductive." For example, Herschel (1830: 164) emphasized that we must not be too "scrupulous as to how we reach knowledge of such general facts: provided only we verify them carefully once detected." And, as Ghiselin (1969) has long argued, Darwin's method can also be construed as being hypothetico-deductive. However, these figures would have shuddered at hearing their views characterized as "hypothetical" or "deductive.' These words were highly suspect at the time. It would be akin today to using the word "Oriental" instead of "Asian" or "Down syndrome" instead of "twenty-one trisomic." At least I am able to refer to John Stuart Mill as a "philosopher of science" because the appellation was common at the time.

Herschel's *Preliminary Discourse* was widely read and favorably reviewed.[2] Herschel pitched his book at just the right level to be read and appreciated by the newly emerging scientists of his day, and once again, I can use the term "scientist" without fear of abuse because Whewell coined the term in a paper he presented at the 1833 meeting of the British Association for the Advancement of Science and used it again in a review of a book on physics written by Mary Somerville (Whewell 1834: 59). However, Whewell coined the term "scientist" only to reject it as being unpalatable. It caught on anyway. According to one semantic thesis, it follows that no scientists existed before 1833. They sprang into existence in the English-speaking world (possibly the world over) only when Whewell coined and rejected the term.[3]

Darwin began his career in science primarily as a geologist and was strongly influenced by Lyell's *Principles of Geology* (1830–33). In a sense, all Darwin did was to extend Lyell's uniformitarianism to include the living world; but as Lyell complained later, if species evolve, he would have to rewrite totally the second volume of his magnum opus (for a detailed discussion of the origin of Darwin's views in Lyell, see Hodge 1983; for Lyell's views on species, see Wilson 1970). Today we break down Lyell's philosophy into three theses: actualism (the only kinds of causes that can be postulated for the past are those in operation today), uniformitarianism (these causes acted at the same rate in the past as they do now), and steady-stateism (the universe exists in a steady state, exhibiting no directional change). Herschel accepted Lyell's first two theses but was a bit reticent about the third, while Whewell rejected the first two and accepted the third (Ruse 1976a).

In spite of Whewell's great reputation for erudition, his philosophical views left his fellow Brits confused. They had great difficulty in making anything out of such notions as "fundamental ideas" which through time could come to be seen as necessary. For nononsense Victorians, Whewell's system sounded too "German," too "idealistic."[4] To this day, Whewell scholars find themselves in deep disagreement over what Whewell really meant to be saying. If scholars who spend their entire careers studying Whewell cannot come to any sort of consensus on the basics of Whewell's philosophy, then perhaps his contemporaries can be excused for their inability to understand it. At least, we know that Darwin read Whewell's *History of the Inductive Sciences* (1837) and his *The Plurality of Worlds* (1853) as well as Herschel's (1841) review of Whewell (Ruse 1977:263).

MILL ON DARWIN

On one point Victorian scholars are in agreement. It was Mill's *System of Logic* that swept the day. Even Whewell (1849: 4) had to acknowledge ruefully that Mill's conception of science had already attained "extensive circulation and numerous, fervent admirers." Thus, it would seem that the view of science with which Darwin had to contend was primarily that of Mill. What influence did Mill's work on the nature of science and scientific reasoning have on Darwin, and vice versa? In the first seven volumes of the monumental *Correspondence of Charles Darwin* (1985–1993; hereinafter, the CCD), Mill is never mentioned. This apparent lack of influence is in itself strange. From 1843 on, right when Darwin was working on his species book, learned journals were full of the debate between Whewell and Mill over proper scientific method. If Darwin read anything on this dispute, he did not say so. His views about proper scientific method were certainly influenced by Herschel and Lyell and, to a lesser extent, by Whewell, but I have not been able to find any evidence that Mill had any direct influence on Darwin prior to 1859. Later Darwin did refer to Mill's work but it was to his utilitarianism and the greatest good principle (Darwin 1871, vol. 1: 71, 97; vol. 2: 328; Darwin 1903, vol. 1: 327–8), not to his principles of scientific method.

However, after the appearance of the *Origin*, Mill was brought to Darwin's attention by Henry Fawcett. After reading so much about how "unphilosophical" the *Origin* was, Darwin (CCD 8: 514) was happy to read Fawcett's (1860) review objecting to all the philosophical cant about Darwin departing from the "true Baconian method." In a letter to Darwin dated July 6, 1861, Fawcett reassured Darwin:

I was spending an evening last week with my friend Mr. John Stuart Mill and I am sure that you will be pleased to hear from such an authority that he considers that your reasoning throughout [the *Origin*] is in the most exact accordance with the strict principles of Logic. He also says the Method of investigation you have followed is the only one proper to such a subject. (CCD 9: 204)

Prior to this conversation with Fawcett, Mill had written a letter to Alexander Bain (April 11, 1860) lauding Darwin's book:

It far surpasses my expectation. Though he cannot be said to have proved the truth of his doctrine, he does seem to have proved that it *may* be true which I take to be as great a triumph as knowledge & ingenuity could possibly achieve on such a question. Certainly nothing can be at first sight more entirely unplausible than his theory & yet after beginning by thinking it impossible, one arrives at something like an actual belief in it, & one certainly does not relapse into complete disbelief. (Mineka and Lindley 1972, vol. 15: 695)

Eight years later (Dec. 2, 1868), in a letter to Herbert Spencer, Mill gives his estimation of Spencer's version of evolutionary theory:

. . . altogether apart from the consideration of what portion of your conclusions, or indeed of your scientific premises, have yet been brought into the domain of proved truth, the time had exactly come when one of the greatest services that could be rendered to knowledge was to start from those premises, simply as a matter of hypothesis, and see how far they will go to form a possible explanation of the concrete parts of organization and life. That they should go so far as they do, fills me with wonder; and I do not doubt that your book, like Darwin's, will form an era in thought in its particular subject, whatever be the scientific verdict ultimately pronounced on its conclusions; of which my knowledge of the subject matter does not qualify me to judge. (Mineka and Lindley 1972, vol. 15: 1505)

Again, in 1869, Mill wrote to Hewett Cottrell Watson (1804–1881):

In regard to the Darwinian hypothesis, I occupy nearly the same position as you do. Darwin has found (to speak Newtonianly) a *vera causa,* and has shewn that it is capable of accounting for vastly more than had been supposed: beyond that, it is but the indication of what may have been, though is not proved to be, the origin of the organic world we now see. I do not think it an objection that it does not, even hypothetically, resolve the question of the first origin of life: any more than it is an objection to chemistry that it cannot analyze beyond a certain number of simple or elementary substances.

Your remark that the development theory naturally leads to convergences as well as divergences is just, striking & as far as I know, has not been made before. But does not this very fact resolve one of your difficulties, viz. that species are not by divergence, multiplied to infinity? since the variety is kept down by frequent blending. The difficulty is also met by the fact that the law of natural selection must cause all forms to perish except those which are superior to others in power of keeping themselves alive in some circumstances actually realized on earth (Mineka and Lindley 1972, vol. 15: 1553–54).

All the preceding quotations are taken from private correspondence. What about Mill's public pronouncements? In later editions of his *System of Logic* (1872: 328), Mill added a footnote:

Mr. Darwin's remarkable speculation on the Origin of Species is another impeachable example of a legitimate hypothesis. What he terms "natural selection" is not only a *vera causa,* but one proved to be capable of producing effects of the same kind with those which the hypothesis ascribes to it: the question of possibility is entirely one of degree. It is unreasonable to accuse Mr. Darwin (as has been done) of violating the rules of Induction. The rules of Induction are concerned with the conditions of Proof. Mr. Darwin has never pretended that his doctrine was proved. He was not bound by the rules of Induction, but by those of Hypothesis. And these last have seldom been more completely fulfilled. He has opened a path of inquiry full of promise, the results of which none can foresee. And is it not a wonderful feat of scientific knowledge and ingenuity to have rendered so bold a suggestion, which the first impulse of every one was to reject at once, admissible and discussible, even as a conjecture?

On the surface, Mill's endorsement of Darwin's theory seems reassuring, at least from our present-day perspective. Natural selection is a *vera causa,* and Darwin used the method appropriate to his investigations – the method of hypothesis. On the surface, Mill seemed to be supporting Darwin, and since surface readings tend to be a good deal more effective in the course of human events than deeper proddings, Mill's contemporaries did see him as supporting

Darwin. However, we philosophers are less interested in what effect a particular thinker had on his or her contemporaries and more interested in what he or she really said. On a deeper reading of Mill's ideas on scientific method, Mill was anything but supportive of Darwin. According to Mill, Darwin had produced a very promising hypothesis, but he had proved nothing.

As in the case of his positivistic descendants, Mill distinguished between the logic of discovery and the logic of proof, and the principle of hypothesis belonged to the logic of discovery, not proof. Mill, however, did not treat this distinction as being as fundamental as his positivist descendants would. He had only just stumbled upon it in his debates with Whewell. In any case, if Mill had thought that *all* scientific theories of his day were only promising hypotheses and none had been proved, according to his standard of proof, I could stop here. Perhaps Mill just had standards of proof so rigorous that no scientific theory can meet them. Such unrealistically high epistemological standards in philosophy are not unusual. For example, generations of skeptics have argued that no one ever really knows anything. Mill did have very high standards of proof. His tables of induction were designed to make the resulting hypotheses as certain as possible. But he also thought that some scientific theories, usually in physics, had been proved. Thus, Mill was making a distinction between certain theories that had been proved according to his standards of proof and others that had not. Darwin's theory of evolution belonged to the second group.

THE PRINCIPLE OF EXCLUSION

The traditional problem of induction concerns moving from the level of *particulars* to the level of *kinds*. One observes that this crow is black, and this crow is black, and this crow is black. Eventually one concludes that all crows are black. One problem with this inference is that too many crows exist for anyone to observe all of them. In addition, huge numbers of crows are dead, and just as many have yet to be born. Even creationists, who believe that a finite number of crows exist, can never hope to observe all crows. Another problem with reasoning from particulars to kinds, as Whewell (1849: 44) complained, is that Mill's inductive tables "take for granted, the very

thing which is most difficult to discover, the reduction of the phenomena to formulae." In reasoning from observations about particular crows to all crows, the concept of crow is being assumed, and that
is the problem to begin with.

According to Mill, however, the most determinate inductive
method was the method of exclusion. If only A, B, or C can cause D,
and neither A nor B is present, then C caused D. In the preceding
statement, A, B, C, and D all refer to kinds of events. Thus, the
traditional problem of induction does not apply. The method of exclusion does not work for particulars because for any natural kind
too many particulars exist. In fact, the number may be not only large
but also indefinite. However, once the level of kinds is reached,
Herschel and Mill thought that the method of exclusion should
work. Many fewer kinds exist than particulars, and their number is,
so Herschel and Mill argued, determinate. If one can list all possible
causes of an event and eliminate all these causes save one, then that
one must be the cause. The problem is listing and eliminating all the
alternative causes (for further discussion, see Hodge, this volume).

All of the preceding reflects the abstract concerns of philosophers.
What about putting the principle of exclusion into practice? After
returning from his voyage on the *Beagle*, Darwin in 1838 visited Glen
Roy in Scotland. Around the sides of this valley, three parallel
shelves could be found. They were obviously shores of former
bodies of water – either lakes or arms of the sea. If these shores were
caused by the successive lowering of the level of a lake, then a huge
dam of some sort must have existed at the mouth of the glen. Since
Darwin could find no remnants of such a barrier, he rejected the lake
hypothesis. This left only arms of the sea. If these shores were caused
by arms of the sea, then either the sea level had dropped successively
through time or the land had been elevated. On the basis of the
available evidence, Darwin concluded that the sea level had not
dropped during this period. Hence, the land must have been elevated (Darwin 1839).

Unfortunately, Agassiz's glacier theory was soon brought to Darwin's attention. A glacier could provide the necessary dam and
disappear (like an icicle used as a knife) leaving hardly a trace. But
when this alternative explanation was presented to Darwin, he did
not give up immediately. Instead, he tried to find a way around this

unanticipated alternative. However, after twenty years of finagling, Darwin (1899, vol. 1: 57) finally was forced to admit that his theory had been "one long gigantic blunder from beginning to end." What lesson did Darwin's gigantic blunder about the parallel roads of Glen Roy teach Darwin? As he himself put it, his error had been a lesson "never to trust in science to the principle of exclusion"! Thus, Darwin rejected the very method that Herschel and Mill were touting as being fundamental to science.

To repeat, if Mill had thought that no scientific hypotheses had ever been proven, then his comments about Darwin's theory would have placed Darwin's theory in the same boat with all other scientific theories, including those of Newton. But Mill thought that certain physical theories had met his standards of proof – theories, I might whiggishly add, that we now take to be seriously defective. Today Darwin's theory holds at least as secure a position in science as those theories that Mill judged vastly superior to it.

Thus, Mill turns out not to be such an unreserved champion of Darwin's theory after all. Unlike Copernicus, Kepler, and Newton, Darwin had proved nothing. But at least Mill thought that evolution by natural selection was a promising hypothesis that might *eventually* be proved. However, before his death in 1873, Mill concluded otherwise. In his letter of 1860 to Alexander Bain, Mill observed that once one had arrived at something like actual belief in Darwin's theory, one certainly did not relapse into complete disbelief. Unfortunately, in the final years of his life, Mill relapsed into just such a state of disbelief. In a paper published posthumously, Mill (1874: 174) presented his final estimation of Darwin's theory, concluding that in the "present state of our knowledge, the adaptations in Nature afford a large balance of probability in favour of creation by intelligence." On the standards of proof that Mill himself helped to develop, Mill is forced to conclude that creation by intelligence is much more probable than evolution through natural selection (see also Hodge, this volume).

RATIONAL EXTRAPOLATIONS

Those casual readers who thought initially that Mill supported Darwin's theory were now forced to conclude that Mill rejected it. If we are to take Mill at his word, one reason that he rejected Darwin's theory was the presence of design. Only an intelligent creative agent could have produced all the marvelous adaptations that we see all around us in the living world. But I think that Mill may have had another reason for rejecting Darwin's theory – though I have yet to find him discussing it. Even if Mill did not see this implication of Darwin's theory as a reason for rejecting it, he *should* have. What follows is the sort of rational reconstruction that philosophers fancy and historians and social scientists despise. In the following discussion, I do not deal with what Mill actually said but with logical conclusions drawn from what he said – whether or not he himself ever noticed these consequences.

According to Mill, species of plants and animals are paradigm natural kinds. And for Mill (1872: 80, 81, 471), natural kinds are classes:

. . . distinguished by unknown multitudes of properties, and not solely by a few determinate ones – which are parted off from one another by an unfathomable chasm, instead of a mere ordinary ditch with a visible bottom. [Real kinds are] distinguished from all other classes by an indeterminate multitude of properties not derivable from one another.

Kinds are Classes between which there is an impassable barrier.

Mill (1872: 440) was aware that zoologists took species to consist of "individuals which have, or may have sprung from the same parents," but he goes on to add that "it is presumed that individuals so related resemble each other more than those which are excluded by such a definition." If genealogy and similarity always went together, it would not matter which criterion one used; but they do not. But most importantly, if species evolve gradually, then there are no impassable barriers between them, and in the absence of such barriers, species as separate entities do not exist. For example, Lyell (1832, vol. 2: 36) acknowledged that a "belief in the reality of species is not inconsistent with the idea of a considerable degree of variability in the specific character." However, if species gradate imperceptibly into each other, as Darwin claimed, then species are not real (Wilson 1970: 88–92, 116, 125, 148, 185, 222, 250–8, 415).

Whewell was also aware that organisms cannot be partitioned neatly into separate natural classes on the basis of the characters that they exhibit. In response to this variation, Whewell introduced his Type Method, according to which a systematist must pick one species as "typical" and classify other species in relation to this type species. For Whewell, species formed a pyramid of increasingly inclusive classes. All that Whewell acknowledged was that the boundaries between these groups of organisms, though not perfectly gradual, are periodically fuzzy. Though fuzzy boundaries are compatible with the gradual evolution of species, they do not necessitate it. For Whewell (1840, vol. 1: 458) the immutability of species was not a generalization from experience but a necessary prerequisite for knowledge:

. . . our persuasion that there must needs be characteristic marks by which things can be defined in words, is founded on the assumption of the *necessary possibility of reasoning* [emphasis in the original].

If species evolve gradually, then even Whewell's Type Method is incapable of handling them. There can be no "characteristic marks" in even a statistical sense. Thus, on Whewell's standards, if species evolve, we cannot reason about them – quite a drawback for the biologists of the day (see Ruse 1976b: 251 for a different reading of these passages).

Of greater importance, Mill took statements about species to be laws of nature. Thus, it follows that if species evolve, laws of nature are evolving! At the time, the immutability of laws of nature went hand in hand with the immutability of such natural kinds as species (Herschel 1830: 39–40; Whewell 1835: 422, 448). If species evolve, then laws of nature evolve, and if laws of nature evolve, then all is lost. We can never know anything for certain. Just because Newton's laws apply today does not mean that they applied in the past or will apply in the future. Thus, contrary to Lyell's uniformitarianism, we must constantly reevaluate our beliefs about nature.

All of the preceding concerns implications that follow logically from various principles that the protagonists held, whether or not these scientists themselves actually were aware of these implications. Other characteristics of Darwin's theory, however, received substantial and explicit attention. For example, many of Darwin's contemporaries were repulsed by his theory because it was so competitive, individualistic, and elitist. Although Darwin's theory did not portray

nature as being totally competitive (he acknowledged some coopera-
tion), individualistic (he entertained the possibility that we now term
species selection), or elitist (the "best" did not always survive to
reproduce), it certainly leaned strongly in these directions. Some of
Darwin's contemporaries were willing to accept such an offensive
view of nature, but most were not. Before most of Darwin's contemporaries could be brought to accept the evolution of species, the
process that produced this transmutation had to be made more pala-
table. Blind variation and selective retention had to be replaced by
directed evolution – evolution had to be in some sense "progressive"
(Ruse 1996) – and this evolution had to occur in leaps, not gradually.
Finally, man had to be excepted. For example, Herschel added a
footnote to the 1867 edition of his *Physical Geography of the Globe*
stating that:

> We can no more accept the principle of arbitrary and casual variation and
> natural selection as a sufficient account, *per se*, of the past and present
> organic world, than we can receive the Laputan method of composing books
> (pushed *a l'outrance*) as a sufficient one of Shakespeare and the Principia.
> Equally in either case, an intelligence, guided by a purpose, must be con-
> tinually in action to bias the directions of the steps of change – to regulate
> their amount – to limit their divergence – and to continue them in a definite
> course. We do not believe that Mr. Darwin means to deny the necessity of
> such intelligent direction. But it does not, so far as we can see, enter into the
> formula of his law; and without it we are unable to conceive how the law can
> have led to the results. On the other hand, we do not mean to deny that such
> intelligence may act according to a law (that is to say, on a preconceived and
> definite plan). Such law, stated in words, would be no other than the actual
> observed law of organic succession; or one more general, taking that form
> when applied to our own planet, and including all the links of the chain
> which have disappeared. But the one law is a necessary supplement to the
> other, and ought, in all logical propriety, to form a part of its enunciation.
> Granting this, and with some demur as to the genesis of man, we are far from
> disposed to repudiate the view taken of this mysterious subject in Mr. Dar-
> win's work. (Herschel 1867: 12)

CONCLUSION

In the early days of philosophy of science in Great Britain, those
scholars engaged in what we think of as philosophy of science were
confident that the feedback between philosophy and science would
be salutary for both. The effects of this interplay with respect to

Darwin were, to say the least, mixed. Herschel, Lyell, Whewell, and Mill should have been in an ideal place to evaluate Darwin's theory. Whewell rejected it outright and never wavered in his opposition. Herschel and Lyell were far from disposed to repudiate Darwin's theory as long as it was reworked to eliminate its more offensive tenets – precisely those tenets that Darwin found most important to his theory. Darwin thought that if he acknowledged the necessity of the continued intervention of creative power, his theory of natural selection would be "valueless" (CCD 7: 354). Finally, Mill began by finding Darwin's theory a promising hypothesis but eventually rejected it for design.

Today many of us engaged in science studies still harbor the belief that scientists would pursue their activities more ably if only they studied a little philosophy of science, and conversely, that philosophers of science would perform their jobs better if only we had more firsthand knowledge of science. I continue to harbor this belief even though the case of Darwin's theory and nineteenth-century British philosophers of science does not exactly support it.

NOTES

1. Both Mill (1831) and Whewell (1831a) reviewed Herschel's (1830) *Preliminary Discourse*. Herschel (1841) in turn reviewed Whewell's (1837) *Philosophy of the Inductive Sciences*, while Whewell (1831b, 1832, 1835) repeatedly reviewed Lyell's *Principles of Geology* and published a book in response to Mill's criticisms (Whewell 1849). For a review of the reviews of Herschel's *Discourse* occasioned by the 1987 reprint of this influential book, see Yeo (1989).
2. Although Mill (1831) was laudatory in his review of Herschel's (1830) *Preliminary Discourse*, he remarked in his autobiography (Mill 1873: 146) that he had read it "with little profit." The same cannot be said for his reading of Whewell's (1837) *History*. Mill garnered the little he knew about science largely from reading Whewell's *History of the Inductive Sciences*.
3. The relevance of various language communities to this semantic thesis remains undetermined. Before 1833 the term "science" was prevalent in English, but not "scientist." After Whewell coined the term "scientist" in English as a possible replacement for "natural philosopher," we are allowed to use the term to refer to scientists as scientists in the English-speaking world, but what about other language communities? Are we

also allowed to term German and French scientists "scientists"? But scientists are not termed "scientists" in French and German.

4. As Yeo (1979: 500) characterized Whewell's fundamental ideas, they were "inherent rather than innate in the mind; while not derived from experience, they did require experience to unfold them; but they also served to organize experience. They were not objects of thought, but the laws of thought. The fundamental ideas were the mind's contributions to the act of cognition."

REFERENCES

Darwin, C. 1839. "Observations on the Parallel Roads of Glen Roy, and of Other Parts of Locheber in Scotland, with an Attempt to Prove that They Are of Marine Origin." *Philosophical Transactions, Royal Society of London* 129: 39–82.

Darwin, C. [1859] 1966. *On the Origin of Species by Charles Darwin: A Facsimile of the First Edition.* Cambridge: Harvard University Press.

Darwin, C. [1871] 1981. *The Descent of Man, and Selection in Relation to Sex.* Princeton: Princeton University Press.

Darwin, C. 1985–1993. *The Correspondence of Charles Darwin.* Cambridge: Cambridge University Press.

Darwin, F., ed. 1899. *The Life and Letters of Charles Darwin,* New York: Appleton.

Darwin, F. 1903. *More Letters of Charles Darwin,* New York: Appleton.

Engels, Eve-Marie. 1995. *Die Rezeption von Evolutionstheorien im 19, Jahrhundert: Herausgegeben, eingeleitet und mit einer Auswahlbibliographie versehen von Eve-Marie Engels.* Frankfurt am Main: Suhrkamp.

Fawcett, H. 1860. "A Popular Exposition of Mr. Darwin's On the Origin of Species." *Macmillan's Magazine* 3: 81–92.

Ghiselin, M. T. 1969. *The Triumph of the Darwinian Method.* Berkeley: The University of California Press.

Herschel, J. F. [1830] 1987. *Preliminary Discourse on the Study of Natural Philosophy: A Facsimile of the First Edition.* Chicago: University of Chicago Press.

Herschel, J. F. 1841. Review of Whewell's *History of the Inductive Science from the Earliest to the Present Times* (1837) and *The Philosophy of the Inductive Sciences Founded upon their History* (1840), *Quarterly Review* 68:177–238.

Herschel, J. F. 1867. *Physical Geography of the Globe.* (4th ed.) Edinburgh: Adam and Charles Black.

Hodge, M. J. S. 1983. "Darwin and the Laws of the Animate Part of the Terrestrial System (1835–1837): On the Lyellian Origins of His Zoonomical Explanatory Program." *Studies in the History of Biology* 6: 1–106.

Hull, D. L. 1972. "Darwin and 19th Century Philosophers of Science." In R. Giere and R. Westfall, eds., *The Foundations of Scientific Method: The Nineteenth Century,* 115–132. Bloomington: Indiana University Press.

Hull, D. L., P. Tessner, and A. Diamond. 1978. "Planck's Principle." *Science* 202: 717–723.

Hull, D. L. 1985. "Darwinism as a Historical Entity: A Historiographic Proposal." In D. Kohn, ed., *The Darwinian Heritage*, 773–812. Princeton: Princeton University Press.

Laudan, L. 1977. *Progress and Its Problems: Towards a Theory of Scientific Growth.* Berkeley: University of California Press.

Lyell, C. [1830–33] 1990. *Principles of Geology: Facsimile of the First Edition in Three Volumes.* Chicago: University of Chicago Press.

Mill, J. S. 1831. Review of Herschel's *Preliminary Discourse* (1830). *Literary Examiner*, March 20, pp. 179–180.

Mill, J. S. 1843. *A System of Logic, Ratiocinative and Inductive, Being a Connected View of the Principles of Evidence, and the Methods of Scientific Investigation.* London: Longmans, Green.

Mill, J. S. 1872. *A System of Logic.* (8th ed.) London: Longmans, Green.

Mill, J. S. 1873. *The Autobiography of John Stuart Mill.* London: Longmans, Green.

Mill, J. S. 1874. *Three Essays on Religion: Nature, the Utility of Religion, and Theism.* London: Longmans, Green.

Mineka, F. E., and D. N. Lindley, eds. 1972. *The Later Letters of John Stuart Mill.* (17 vols.) London: Routledge & Kegan Paul.

Ruse, M. 1975. "Darwin's Debt to Philosophy: An Examination of the Influence of the Philosophical Ideas of John F. W. Herschel and William Whewell on the Development of Charles Darwin's Theory of Evolution." *Studies in the History and Philosophy of Science* 6: 159–181.

Ruse, M. 1976a. "Charles Lyell and the Philosophers of Science." *British Journal for the History of Science* 11: 121–131.

Ruse, M. 1976b. "The Scientific Methodology of William Whewell." *Centaurus* 20: 227–257.

Ruse, M. 1977. "William Whewell and the Argument from Design." *The Monist* 60: 244–268.

Ruse, M. 1996. *Monad to Man: The Concept of Progress in Evolutionary Biology.* Cambridge: Harvard University Press.

Whewell, W. 1831a. Review of Herschel's *Preliminary Discourse* (1830). *Quarterly Review of London* 45: 374–377.

Whewell, W. 1831b. Review of volume 1 of Lyell's *Principles of Geology. The British Critic, Quarterly Theological Review and Ecclesiastical Record* 9: 180–206.

Whewell, W. 1832. Review of volume 2 of Lyell's *Principles of Geology* (1832). *Quarterly Review of London* 47: 103–132.

Whewell, W. 1834. Review of Mary Somerville's *On the Connexion of the Physical Sciences* (1834). *Quarterly Review* 51: 54–68.

Whewell, W. 1835. Review of Lyell's *Principles of Geology* (3rd ed.). *Quarterly Review of London* 53: 406–448.

Whewell, W. 1837. *History of the Inductive Sciences.* London: Parker.

Whewell, W. 1840. *The Philosophy of the Inductive Sciences, Founded upon Their History.* London: Parker.

Whewell, W. 1849. *Of Induction, with Especial Reference to Mr. J. Stuart Mill's System of Logic.* London: Parker.

Whewell, W. 1953. *On the Plurality of Worlds,* London: Parker.

Wilson, L. G., ed. 1970. *Sir Charles Lyell's Scientific Journals on the Species Question.* New Haven: Yale University Press.

Yeo, R. 1979. "William Whewell, Natural Theology and the Philosophy of Science in Mid Nineteenth Century Britain." *Annals of Science* 36: 493–516.

Yeo, R. 1989. Review of Herschel's *Discourse. Studies in the History and Philosophy of Science* 20: 541–552.

Chapter 4

The Epistemology of Historical Interpretation

Progressivity and Recapitulation in Darwin's Theory

ROBERT J. RICHARDS

Historians, philosophers, and scientists usually read documents of past science differently. The scientist, most often, is curious about the ways in which events, experiments, observations, or theories described in the documents made clear the path, or blocked it, to the contemporary scientific state. Philosophers, while interested in the developmental history of contemporary science, usually focus on the validity of theories and the ways in which experiments or observations affected that validity. Historians do not neglect either the developmental trajectories of science or questions of validity. They try, however, to examine the trajectories from the founding ground rather than simply retracing the sources of contemporary science in retrospect; and they seek to determine why the originators or their opponents regarded the science in question as valid or invalid. The historian is constantly discovering the nasty little facts that make of short life some beautiful construction of the scientist or philosopher. To be sure, the historian must be familiar with the end point of a particular developmental sequence. The historian of evolutionary theory, for instance, needs to know a good deal about contemporary evolutionary concepts in order to pick out from the past those antecedents that must have given rise to the recent ideas. But chastened by the cardinal precept of the discipline, the historian will guard against the seductive substitution of more recent theory for older theory, against the assumption that the child must be a clone of the parent. The historian will understand that the commonplaces of today's science, even if perfectly "logical" extensions of the originating theory, may simply not have been recognized by older scientists. Imre Lakatos once suggested that even if Bohr had not thought of electron spin in 1913, the historian "describing with hindsight the Bohrian programme, should include electron spin in it, since electron spin fits naturally in the original outline of the programme."[1] Though Lakatos may have made this suggestion with a Hungarian twinkle in his eye, it yet reveals the pernicious metaphysics underlying an epistemological assumption usually made by philosophers and scientists (and, alas, by some historians) when attempting to recover past science. They tend to regard scientific theories as timeless, abstract entities, instead of historical creatures. Perhaps no philosopher has displayed the desiccated metaphysics of this assumption more prominently than Karl Popper, who argued that scientific theories, as well as other cultural artifacts, do live in a Platonic third world, a timeless realm from which they occasionally come aborning into history, trailing clouds of immortality.[2] Comparable metaphysical entities were banished from biology with the publication of Darwin's *Origin of Species.* It is time they were exorcized from the epistemology of historical interpretation as well.

In this chapter, I wish to explore, in a concrete way, just how the assumption of the nonhistorical nature of theories can derange our construction of past science and, consequently, mislead us in our understanding of contemporary science. The case is the historical commonplace that Darwin's theory of evolution by natural selection was not a progressive conception. At the end of this essay, I will suggest a view of theories that, I believe, protects against the liabilities here diagnosed.

THE STATUS QUO ANTE IN THE HISTORIOGRAPHY OF EVOLUTIONARY PROGRESSIVISM

Until recently, virtually all scientists, philosophers, and historians who analyzed Darwin's theory argued that it denied any notions of progress or hierarchy as the result of evolutionary processes. Stephen Jay Gould, for example, has maintained that "an explicit denial of innate progression is the most characteristic feature separating Darwin's theory of natural selection from other nineteenth century evolutionary theories. Natural selection speaks only of adaptation to local environments, not of directed trends or inherent improvement."[3] Like Lakatos, Gould urged that the logic of Darwin's theory denied the possibility of evolutionary progress. Gould observed that in a shifting environment – the usual assumption – new adaptations would have to be acquired that might change entirely the previous direction of trait acquisition. So Darwin's theory could, indeed, be distinguished from other evolutionary theories of the late nineteenth century, those of Spencer and Haeckel, for instance. Theories of this non-Darwinian vintage clearly did imply a steady, progressive advancement of life on this Earth.

Gould has been one of the very strong opponents of notions of evolutionary progress, holding that such ideas bespeak more politics than science. He has argued, following this thread, that the "Victorian unpopularity" of Darwin's theory stemmed from its destruction of the foundations for progressivist notions in the political sphere: "It [evolution by natural selection] proposes no perfecting principles, no guarantee of general improvement; in short, no reason for general approbation in a political climate favoring innate progress in nature."[4] Peter Bowler, an historian, is another who endorsed the common view. Bowler has maintained that "Darwin's mechanism challenged the most fundamental values of the Victorian era, by making natural development an essentially haphazard and undirectional process."[5]

THE PRELIMINARY CASE FOR DARWINIAN PROGRESSIVISM

Those who argued that Darwin's theory was not progressivist in character had initially to dispatch or, more commonly, to ignore some rather straightforward statements in the *Origin of Species* that seemed to make sweeping progressivist claims. For example, in the penultimate paragraph of the first edition of the book, in summing up his position, Darwin declared that "as natural selection works solely by and for the good of each being, all corporeal and mental endowments will tend to progress towards perfection."[6] Darwin's conclusion about the progressive effects of natural selection directly flowed from the logic of his device *as he understood it*. In January of 1839, in his Notebook E, after he had come self-consciously to the formulation of natural selection, Darwin began to construct his device in a dynamically progressivist way. He supposed that the environment against which organisms would most often be selected would be the living environment of other creatures. So, as some creatures constituting that living environment improved, became more complex and better adapted, others would also have to improve or go to the wall. Progressive adaptations would beget progressive adaptations. Thus reciprocal developmental responses would be evoked throughout the living system of organisms. Darwin certainly recognized that the extinction of simpler animals in a particular location might lead to some more advanced creatures themselves becoming simplified – to backfill the gap. This would mean that some animal series might show a kind of devolution; but the general trend would, nonetheless, be inevitably toward ever-increasing complexity and improvement. He put it this way in his notebook:

The enormous number of animals in the world depends on their varied structure & complexity. – hence as the forms became complicated, they opened fresh means of adding to their complexity. – but yet there is no necessary tendency in the simple animals to become complicated although all perhaps will have done so from the new relations caused by the advancing complexity of others. – It may be said, why should there not be at any time as many species tending to dis-development (some probably always have done so, as the simplest fish), my answer is because, if we begin with the simpler forms & suppose them to have changed, their very changes tend to give rise to others.[7]

Darwin rejected what he assumed to be Lamarck's belief in an intrinsic cause of necessary progress buried in the interstices of animal organization. For Darwin, progress resulted from an extrinsic causal relationship, namely that of natural selection, which would inevitably produce progressive changes throughout most lineages.

In the *Origin*, Darwin integrated this conception of natural selection's progressive dynamic into his more general theory of divergent evolution. In large, open areas the environment of other closely related species would promote mutually adaptive lineages, the species of which would, as a result, continuously diverge and improve. Darwin echoed the formulation of his notebook, expressing it this way in the *Origin:*

the conditions of life are infinitely complex from the large number of already existing species; and if some of these many species become modified and improved, others will have to be improved in a corresponding degree or they will be exterminated.[8]

The general improvements created by natural selection, Darwin maintained elsewhere in the *Origin,* would not be merely relative to local environments. Now, most scholars have urged that Darwin's theory allowed only local improvements at best. The historical Darwin, though, was of a different mind. He thought natural selection would produce ever more progressive types over time, so that "the more recent forms must, on my theory, be higher than the more ancient; for each new species is formed by having had some advantage in the struggle for life over other and preceding forms."[9] As a kind of thought experiment, Darwin considered what would happen were older fauna to struggle with more recent types in neutral environments. Without a thought to the claims of people like Gould or Bowler, he concluded that eocene fauna would inevitably succumb to modern types, just as secondary fauna would be beaten by eocene and paleozoic by secondary.[10]

To modern eyes, Darwin's standard for measuring progress might seem a bit vague. He was himself sensitive to the difficulties. In his Notebook B, in 1837, he remarked that "[i]t is absurd to talk of one animal being higher than another"[11] – a line often quoted to indicate Darwin's nonprogressivist stance. At this early stage in his thinking, what Darwin really objected to, though, was any talk of progress that remained unanchored in a natural system of classification. As he suggested to his friend George Waterhouse, the great entomologist, only within a natural system did designations of higher and lower types make sense.[12] Of course, Darwin believed he had just such a system. In light of his natural system of descent, he applied a set of three necessary and together sufficient criteria for applying to organisms the epithets "higher" and "lower." The first criterion was complexity of organization, which even he recognized as not very sharply defined. But in practice, Darwin did not hesitate to point out examples in which complexity served as a marker of perfection and progress – the vertebrate eye, to Darwin's mind, was obviously such a case.[13] The second criterion, which Darwin employed throughout the *Origin,* did help to specify the first: it was Milne-Edwards's notion of the division of physiological labor, according to which those organisms with greater differentiation of parts would be regarded as more perfect, since they would be able to take advantage of particular corners of a diversified habitat.[14] Earlier, in his study of barnacles, Darwin had invoked a comparable criterion, though attributing it to von Baer. To this criterion he added an interesting, if somewhat circular qualification:

On the whole, I look at a cirripede as being of a low type, which has undergone much morphological differentiation, and which has, in some few lines of structure, arrived at considerable perfection – meaning, by the terms perfection and lowness, some vague resemblance to animals universally considered of a higher rank.[15]

Greater complexity and differentiation would constitute those kinds of general improvements produced by natural selection. The process would be cumulative: natural selection would stack one set of improvements upon another set, as adaptations continued to build in response to ever-changing environments. Where environments had remained stable – not a common occurrence, Darwin believed – there would, however, be little improvement required and stasis might occur; and where gaps appeared in the economy of nature, there might be positive devolution. Yet most lineages would ultimately and necessarily tend toward greater perfection, as Darwin made clear in the third and later editions of the *Origin:*

Natural selection acts, as we have seen, exclusively by the preservation and accumulation of variations, which are beneficial under the organic and inorganic conditions of life to which each creature is at each successive period exposed. The ultimate result will be that each creature will tend to become more and more improved in relation to its conditions of life. This improvement will, I think, inevitably lead to the gradual advancement of the organisation of the greater number of living beings throughout the world.[16]

It is worth pausing a moment to consider what Darwin meant by saying the progressive development of creatures was "inevitable." Philosophers, at least of a more vintage character, like to distinguish three kinds of necessity or inevitability: contingent, physical, and logical. Contingent necessity is trivial: what has already occurred must necessarily be the case. Logical necessity arises from the rules by which we order linguistic propositions. Physical necessity derives from the operations of physical laws, and this seems to be what Darwin meant by the "inevitable" progress of creatures throughout the world. Thus Darwin suggested that all other things being equal and without contingent interfering causes, the laws of biology would ensure that on the whole creatures progressively advance in complexity and, consequently, survivability.

The final criterion of progress, more than the others, marks Darwin as a child of his time. Contemporary critics of the notion of evolutionary progress usually find in the discussions of most advocates in the nineteenth century – say, someone like Ernst Haeckel – an implicit anthropocentric element, that is, an assumption that organisms become more progressively evolved as they approach man, as they become more like us. It has been thought that Darwin escaped including this kind of ideological component in his theory. But clearly he did not. In a note dated March 1845, he reflected:

What is the highest form of any class? Not that which has undergone most changes. For changes may reduce organization: – generally, however, that which has undergone most changes & which approaches nearest to man.[17]

In the third and subsequent editions of the *Origin*, Darwin reiterated this criterion when he observed that in making judgments of advancement in the vertebrates, "the degree of intellect and an approach in structure to man clearly come into play."[18] This anthropocentric evaluation of progress became even more refined when, in the *Descent of Man*, he worried that natural selection might be thrown out of gear, since the Irish – clearly a breed intellectually and morally inferior to the higher English sort – appeared to be propagating at a more rapid rate than their betters. But the Irish, to Darwin's mind, were certainly not an improved type.[19]

RECENT REEVALUATIONS OF DARWIN'S THEORY

During the last decade, some scholars have begun to argue that Darwin's theory, previous historiography notwithstanding, manifests deep veins of progressivism.[20] Even Gould now admits that Darwin's conception had surface dimensions, at least, that appear progressivistic. Gould, however, attributes these features not to the theory of evolution by natural selection, but to Darwin's social pre-dilections.[21] Michael Ruse has recently come to much the same con-clusion. He does so in a grand and rich volume devoted to the estimation of the ways in which progressivist ideology infected vir-tually all evolutionary theories in the nineteenth and early twentieth centuries. Ruse maintains that Darwin incorporated into his theory both "relativistic progress" and "absolute progress" – that is, pro-gress relative to a given environment and progress that leads to a hierarchy of creatures, with human beings in the lead. "Darwin tried to show," according to Ruse, "that we humans – especially Anglo-Saxon humans–are the apotheosis of the evolutionary process."[22] But like Gould, Ruse tempers this historiographic evaluation. He does so in two ways. He first argues that the progressivist strands of the theory – both the "relativistic" and the "absolute," as he denomi-nates them – had virtually no scientific evidence to support them: Darwin's "human-centered progressionism transcended the avail-able data," Ruse claims.[23] "Darwin," he further observes, "did not do any original research of his own, preferring to rely on his – admit-tedly extensive – literature search."[24] But for Ruse, evidence is al-most beside the point, since he maintains that the progressivist traits of Darwin's theory were cultural imports, quite antithetic to the "sci-entific" aspects of the theory:

. . . what was at stake here was Darwin's commitment to the design-like nature of the world, together with his equally strong commitment to a God who works only through unbroken law. But, whether this is legitimate or not, it was a vision that Darwin was imposing on the organic world, not a deduction made by reasoning from the evidence.[25]

According to Ruse, the idea of progress was a cultural value, an ideology, that Darwin infused into his theory. It was an element that accounted for the success of Darwin's publications, but that itself had no real empirical support and could not claim scientific status. It had, by Ruse's lights, no "epistemic" ground. Gould, as previously indicated, regards notions of progress as ideological add-ons to Dar-win's theory. Ruse likewise thinks Darwinian notions of progress to be foreign to the scientific core of the theory – those notions had neither epistemic purchase nor evidentiary validity. Ruse's judg-ments about Darwin's epistemology and the validity of his progressivistic notions are made, perhaps needless to say, in light of our contemporary standards. And digging a bit deeper into the underlying assumption that Ruse has engaged here, one can find, I believe, a certain theory of theories being invoked. He has supposed that Darwin's real theory is an abstract, timeless entity that looks suspiciously like our basic modern, neo-Darwinian theory. And in making this assumption, Ruse has a close philosophical companion in Michael Ghiselin.

Ghiselin has constructed another kind of analysis of Darwin's progressivism. In a hagiographic mode, Ghiselin cannot imagine Darwin's soul being sullied by anything like ideology. He has argued that any progressivist claims that Darwin makes would be in the nature of empirical hypotheses being tested "in light of empirical evidence, experiments, and the like."[26] This view, of course, keeps Darwinian theory respectable – respectable, that is, according to our contemporary notions about scientific probity. And that move is precisely the one that unites Ghiselin's effort with those of Gould and Ruse, despite their differing conclusions about the precise relationship between Darwin's progressivistic assertions and his general theory of natural selection. All three scholars wish to preserve the scientific integrity of Darwin's theory of evolution by natural selection, *according to our contemporary evaluations of science.* And I believe they do so because of a certain conception they harbor concerning the nature of scientific theory. Their epistemology of science (as well as personal predilections) constrains them in their respective judgments about the role of progress in Darwin's theory. I will return to what I take to be their epistemological position in a moment. Suffice it to say at this point, I believe they take scientific theory much in the way of Popper: namely, as a timeless, abstract entity clothed according to the mode of the day, yet capable of donning any guise as the times and fashions change. A theory, according to this view, would, however, willingly stand naked for the loving inspection of those of requisite discernment. I believe, on the contrary, that scientific theories resemble biological organisms: they are historical entities existing within a certain environment, and their structure has been formed both by that environment and by the hereditary legacy of previous theory.

RECAPITULATION AS MANIFESTING THE DEEP STRUCTURE OF DARWINIAN PROGESSIVISM

Perhaps no biologist of the nineteenth century reveals the historicity of scientific theory more than Ernst Haeckel. He became Darwin's champion in Germany, and indeed throughout the world. More people, in fact, learned of Darwin's theory through Haeckel's many publications than through the hand of the master himself.[27] Haeckel's *Die Welträthsel*, which placed evolutionary ideas in a broader philosophical and social context, sold over 100,000 copies in the year of its publication (1899) and some three times that number during the next thirty years – and this only in the German editions.[28] By contrast, during the three decades between 1859 and 1890, Darwin's *Origin of Species* sold only some 39,000 copies in the six English editions.[29] As is well known, Haeckel made recapitulation the cornerstone of his rendition of Darwinian theory. He maintained that during gestation the embryo of a more advanced creature passed through the morphological stages of its adult ancestors. According to this theory, the human embryo, for instance, would develop from a stage at which it initially resembled a primitive one-celled creature, then would exhibit the morphology of an invertebrate, next an early vertebrate, and then would pass through the stage at which it resembled an ancestral apelike creature, finally achieving the human form. Since the embryo would exhibit the adult stages of its ancestors, its ontogeny would provide a picture of the evolutionary history of the phylum. Haeckel's summary coinage that "ontogeny is the short and quick recapitulation of phylogeny" became the watchword for evolutionary theorists at the turn of the century.[30] Undoubtedly what appealed to Haeckel was indeed the recognition that the recapitulational idea pulsed in harmony with his general interpretation of evolution as progressivistic.[31] But what was the case with Darwin? Did his general progressivistic interpretation of evolution recruit recapitulationism? Did the recapitulational idea play any significant role in Darwin's general theory, as it certainly did in Haeckel's?

Most scholars have argued that the idea was actually quite foreign to Darwin's approach. Gould, for instance, has offered a reading of Darwin's *Origin* that neatly places the Englishman's theory on the other side of the European divide.[32] And for Bowler, "recapitulational theory illustrates the non-Darwinian character of Haeckel's evolutionism."[33] The rationale of scholars like Gould and Bowler seems clear enough. They wish to protect authentic evolutionary science from the political taint of progressivism and the more serious danger of having racist ideology bred in the bone, in the very marrow of contemporary evolutionary theory. Gould has argued that Haeckel's progressive evolutionism made the biological path straight for the Nazi panzers to roll through the mid part of this century.[34] But authentic Darwinian theory could not sanction anything so horrendous. After all, under the Platonic epistemology of abstract theories, Darwin's theory is our theory; but our theory stands in opposition to progessivism and, therefore, racism. Consequently, so does Darwin's. Darwin's theory, therefore, at a deep level must be quite distinct from Haeckel's. It must disavow recapitulation.

The actual history of Darwin's theory, however, tells a different story. Recapitulationism has an intrinsic and vital place in the conception that Darwin nurtured from just after the *Beagle* voyage through the last editions of the *Origin of Species* and the *Descent of Man*. His long-term pursuit of the idea stemmed from his conviction that embryological recapitulation supplied the strongest proof for species evolution. Right after the publication of the *Origin*, he responded to Asa Gray, who remarked on the use of embryology in the book. Darwin wrote his friend that "embryology is to me by far the strongest single class of facts in favour of change of form."[35] Even Darwin's colleagues recognized the importance of embryological recapitulation. At the famous Oxford Debate between Samuel Wilberforce and Thomas Henry Huxley, the zoologist and Darwin supporter John Lubbock added the point that "the embryology of the individual in many cases represents the past history of the Species."[36] Recapitulation, for Darwin, was no unimportant idea.

Darwin, of course, was hardly the first biologist to argue for recapitulation. The connection between the two modes of morphologically progressive change – that of species and that of the embryo – had been made initially by German biologists at the beginning of the nineteenth century. David Friedrich Tiedemann, for instance, claimed that "the entire animal kingdom has its evolutionary stages [Entwickelungsperioden], similar to the stages which are expressed in individual organisms."[37] Karl Ernst von Baer and Richard Owen regarded these conceptions of morphological change, in the species and in the individual, as virtually the same, and condemned them as such. For instance, Owen indicted the two ideas in his Hunterian lecture of May 1837 – a lecture that Darwin is very likely to have heard. "The doctrine of transmutation of forms during the Embryonal phases," he asserted, "is closely allied to that still more objectionable one, the transmutation of Species."[38] Darwin also regarded these doctrines as virtually the same, but of course argued strongly for them, in conscious opposition to Owen.

Darwin had long assumed that individual alterations and species alterations were connected. Indeed, in his first musings on species transformation, in the early 1837 passages in the Red Notebook, he compared the process of species change to individual growth.[39] And on the very first page of his first transmutation notebook, Notebook B, he advanced the principle of recapitulation. In the passage, he considered two sorts of propagation, nonsexual budding and splitting, and the ordinary, sexual kind. He wrote:

The ordinary kind, ⟨the⟩ which is a longer process, the new individual passing through several states (typical, ⟨of the⟩ or shortened repetition of what the original molecule has done).[40]

The idea that the embryo passed through morphological stages comparable to that of the "original molecule" in its transformational history – this idea often recurred in Darwin's notebooks and loose notes. So, for instance, again in Notebook B, he considered how layers of progressive adaptations would be laid down in the embryo and preserved:

An originality is given (& power of adaptation) is given by true generation, through means of every step of progressive increase of organization being imitated in the womb, which has been passed through to form that species. – ⟨Man is derived from Monad⟩.[41]

In the some twenty years that separate these passages in his notebooks from the *Origin of Species*, Darwin worked on several principles that made recapitulation a logically necessary concomitant of his general theory of species descent. The first two principles concern the locus of hereditary change.

Darwin argued, from the 1840s right through to the last edition of the *Origin*, that adaptations selected by natural selection or produced through use and disuse would usually occur in the later stages of development. Two facts, he believed, supported the principle: first, that in the womb or in the egg, an embryo did not face differential pressures of natural selection or use-inheritance; second, that only when the mature organism made its way into the world would those pressures be exerted. Hence, adaptational structures would "supervene at a not very early period of life."[42] The second principle had, perhaps, less evidentiary weight, but was no less believed, namely that variations occurring in the adult would recur at the "corresponding age in the offspring."[43] These two principles of evolutionary acquisition of adaptations had three important consequences for Darwin. First, they meant that in very early periods of life's history, parents and their embryos would be more morphologically similar than they are today; but as terminal modifications were added to development in the course of descent, the adult organisms would come to differ substantially from their own embryos. In the *Origin*, Darwin indicated the kinds of experimental study he undertook to demonstrate the second part of this conclusion: newborn puppies of different breeds, for example, showed greater similarity among themselves than with the adults of their own breed.[44] The second and very important consequence of these hereditary principles was that he now had, Darwin thought, an explanation of why embryos of different species within a given class tend morphologically to resemble one another: that was because natural selection or use-inheritance would operate only on the more mature individuals, not the embryos or young. And since the embryos of related species would be touched little by selection, they would tend to preserve the common form of their adult ancestor. Finally, those principles of heredity meant that embryos of contemporary organisms would pass through, in their development, the stages that characterized the evolved forms of their ancient progenitors. Darwin made this last point explicit in his essay of 1844, the first large sketch of his theory:

> . . . it may be argued with much probability that in the earliest & simplest condition of things that the parent & embryo would always resemble each other, & that the passage through embryonic forms is entirely due to ⟨modifications⟩ subsequent variations affecting only the ⟨later states⟩ more mature periods of

life. If so, the embryos of the existing ⟨mammifers⟩ vertebrata will shadow forth the full grown structures of some of the ⟨earliest parent-stocks⟩ forms of this great class, (which existed at the earliest period of the earth's history).[45]

Thus the embryos of contemporary animals would develop through a series of morphological stages that pictured the adult forms of their ancestors. Now this is, of course, exactly the same principle as Haeckel's biogenetic law. In Darwin's view, we would have in the embryos of contemporary organisms a kind of moving daguer-reotype of the history of phylogenetic descent.

Some scholars, like Ghiselin, think it "utterly preposterous" to claim as I have that Darwin believed the embryo offered a kind of moving picture of the history of species descent.[46] The vehemence of Ghiselin's rejection certainly measures commitment at a level be-yond the normal, a commitment to a particular view of theories and much else, I suspect. The evidence already cited would make most scholars yield. In some abnormal instances, though, even Darwin's *ipsae verbae* might not produce the warranted response. Yet on histor-ical issues such as this, the words should matter. This is Darwin from the first edition of the *Origin*:

. . . the adult differs from its embryo owing to variations supervening at a not early age, and being inherited at a corresponding age. This process, whilst it leaves the embryo almost unaltered, continually adds, in the course of successive generations, more and more difference to the adult. Thus the embryo comes to be left as a sort of picture, preserved by nature, of the ancient and less modified condition of each animal.[47]

Darwin, at least, thought the embryo provided us a pictorial review of the evolutionary past, preposterous to modern eyes though it may be.

Darwin continued to reiterate this principle of the embryo re-capitulating the adult stages of the ancestors through all editions of the *Origin*. So, for example, in the last edition (1872), he succinctly stated:

As the embryo often shows us more or less plainly the structure of the less modified and ancient progenitor of the group, we can see why ancient and extinct forms so often resemble in their adult state the embryos of existing species of the same class.[48]

Darwin did not believe, to be sure, that the principle of recapitula-tion held absolutely. There would be one class of possible exceptions to embryological recapitulation: when the embryo was not safely ensconced in the womb or egg but led an independent existence as a larva. Thus exposed to the forces of selection, embryological adaptations might occur; and this would often be the case, Darwin observed, with insects. For the higher animals, however, he seems to have thought there would hardly be any exceptions – at least both his language and logic suggest this. So, for instance, in the sixth edition of the *Origin*, just after considering the exception of insect larvae, he urged:

So again it is probable, from what we know of the embryos of mammals, birds, fishes, and reptiles, that these animals are the modified descendants of some ancient progenitor, which was furnished in its adult state with branchiae, swim-bladder, four fin-like limbs, and a long tail, all fitted for an aquatic life.[49]

I presume that, if pressed, Darwin would have allowed for some few exceptions to exact recapitulation in the higher animals. But, unlike Haeckel, he never developed a thoroughgoing account of the range and character of exceptions to an exact recapitulation.

There can be no reasonable doubt – though, strangely, almost every scholar who has considered the question has doubted it in the most forceful terms – that Darwin, both early and late, held to a strong principle of recapitulation. He knew, of course, that the principle was not perfectly demonstrable from the evidence, since that would have required a comparison with the geological record of fossils, which, as he argued, was a very imperfect record. But he actually reversed the methodological situation: he suggested that recapitulation might serve to confirm the progressive character of the paleontological record. In the *Origin*, he put it this way:

The inhabitants of each successive period in the world's history have beaten their predecessors in the race for life, and are, in so far, higher in the scale of nature; and this may account for the vague yet ill defined sentiment, felt by many palaeontologists, that organisation on the whole has progressed. If it should hereafter be proved that ancient animals resemble to a certain extent the embryos of more recent animals of the same class, the fact will be intelligible.[50]

Owen had thus been correct: belief in the transmutation of forms during the embryonal phase led to a belief in the transmutation of species.

AN ESSENTIALISTIC RENDERING OF DARWIN'S EMBRYOLOGICAL THEORY

I have argued that the tendency of scholars strongly to deny that Darwin advanced a progressivist theory that also endorsed recapitulation can be traced, at least in part, to an epistemological theory about theories, namely that scientific theories are timeless, abstract entities that can be instantiated at different periods while remaining essentially the same. Since much contemporary neo-Darwinism rejects progressivism and recapitulation, so must Darwin's original theory, or so the implicit argument runs. This kind of essentialism about theories is unwarranted by the historical evidence and highly suspect in the face of the empiricism that most scientists and philosophers also usually endorse. This tendency to take the abstract for the historically concrete, though, can be subtle in the work of a scholar. Ghiselin's own account of what appears to be Darwin's sanction of recapitulation presents a nice case of this.

Here is Ghiselin's explanation of what appears to be a strong principle of recapitulation in Darwin's theory:

To explain the connection between ontogeny and phylogeny, Darwin elaborated a sophisticated theory based upon the laws of nature that govern embryological development. A variation can occur at any time in the life cycle of an organism. The same variation will occur at the same time in succeeding generations and may be favored by selection. If variation is purely random with respect to the time that it occurs, then it follows that fewer variations will occur in the earlier stages. Therefore, the earlier stages will be less labile and more likely to retain the ancestral state more or less unchanged. Such a model by no means rules out the occurrence of modifications early in life but only reduces the relative probability that the necessary condition of variation will be met.[51]

This rendition ignores completely what seems embarrassing to modern eyes, namely the vision of the embryo as sequentially picturing the *adult* stages[52] of its evolutionary ancestors – a "preposterous" idea, according to Ghiselin. But let us focus on Ghiselin's reconstruction of Darwin's rationale for believing the embryo remained little modified in the womb. In the above account, Ghiselin offers a probabilistic analysis of variation as modeling Darwin's own. But did Darwin indulge in this kind of probabilistic thinking? I do not believe so. Were his understanding similar to Ghiselin's, then you might expect him also to hold, as I suppose most modern embryologists would, the clear logical consequence of this model of random change. The consequence is this: given that the early stages of embryological development occur only over a modest part of the life of an organism, yet in the course of countless generations the modifications would have accumulated in substantial numbers, it follows that the morphological stages of the embryo would bear only the faintest resemblance to previous forms in the history of the phylum. Such a logical consequence of the random model, however, would provide scant conceptual evidence for recapitulation and no inclination to look upon the embryo as rather faithfully reflecting the history of the phylum. Moreover, Darwin explicitly stated that the modifications by which species gradually attained their structure "supervened at a not very early period of life."[53]

Ghiselin provides a modern probabilistic explanation for the embryo's not suffering much change in the womb. But there is no evidence that Darwin even had the conceptual equipment to think quite in those terms. Rather, Darwin offered three very different kinds of reason for arguing that the embryo remained little changed in the womb and that it only added successive terminal structures during the course of phylogenetic history. First, the embryological environment for most animals would be similar and constant: any variations that arose would have little utility and would be subject to hardly any selective forces; any variations that spontaneously arose would simply not be preserved. Hence, only those variations that initially occurred in the mature forms of the ancestor, the useful ones, would produce those morphological changes that would be retained (not acquired) by the embryo as it passed through the stages of the ancestors.[54] Second, the functions of use and disuse, which supplied not only variations for natural selection to operate upon but also introduced morphological changes – these functional operations would be engaged in differentially only by organisms making their way in the world. Indeed, in the section of the *Origin* where Darwin discusses recapitulation, he mentions only use and disuse as the devices of change, not natural selection.[55] Finally, Darwin simply accepted as an axiom that variations occurring in the adult would recur at a "corresponding age in the offspring."[56] Hence, over a short period, the adult's modifications would be displayed by the offspring only when it became an adult. Over the longer period, the older adult morphology would recede into the terminal stages of embryological development. Thus Darwin's reasons for generally excluding modifications in morphology during embryogenesis had nothing to do with the kind of probability scenario advanced by Ghiselin. Ghiselin's is a bit of thoroughly contemporary reasoning.

The evidence suggests that Darwin had a conclusion he wanted to reach and he marshaled his arguments – and the principles I have just mentioned – in order to reach it. He wanted the embryo to be unaffected in the womb so that he could assert that "the embryo comes to be left as a sort of picture, preserved by nature, of the ancient and less modified condition of each animal."[57] In lieu of a convincing paleontological record – and when Darwin wrote, no such record existed – such an embryological picture of past history would provide concrete, empirical evidence of the transmutation of species. And Darwin believed, as he wrote to Asa Gray and others, that embryology provided the strongest kind of evidence for alteration of species over time.[58]

CONCLUSION: AN HISTORICAL MODEL OF THEORIES

In this historical examination of Darwin's theories of progress and recapitulation, I have tried to show the disadvantages of regarding such theories as timeless, abstract entities. This is the way they have been most usually understood and the way that leads, I believe, to a pernicious presentism in history of science, as I have tried to show in the historical account just elaborated. The temptations to presentism are, however, significantly mitigated if we take theories to be historical entities. Moreover, the abstract model bears the seeds of its own destruction, harboring as it does several infelicities for history of science. By contrast, the historical model of theories has certain positive advantages, which I have suggested in the preceding historical analysis but wish now to specify more carefully.

To regard Darwin's theory, for example, as a timeless, abstract entity leads, first, immediately to the view that the theory, once constituted, remains stable and does not change. But we know that over the course of the more than twenty years he worked on his theory prior to the publication of the *Origin*, Darwin's ideas underwent significant transformation. During the early years of that period, one device of species change replaced another until he hit upon natural selection. And even the device of natural selection underwent modification as Darwin pondered its features: at one time he held natural selection to produce perfect adaptations, but then he altered his view – and so on for the several aspects of the theory.[59] In order to reconcile this historical process with an abstract view of theories, we would have to talk about a new theory each time a fundamental change occurred, suggesting that Darwin had many theories of evolution – a proliferation for which the abstract model is supposed to serve as Occam's razor. Second, one must ask about the relation of Darwin's theory to that of other Darwinians, say the theories of Alfred Russell Wallace, the codiscoverer of evolution by natural selection, or of George Romanes, Darwin's protégé, or of Ernst Haeckel, his great champion in Germany. They were Darwinians, but the theories they held at any one time diverged considerably from Darwin's and from each other (e.g., Wallace did not accept the inheritance of acquired characters, the rest did; their respective notions about what got inherited differed widely, etc.) Certainly all four could not have entertained tokens of the same abstract type, since their respective theories embodied significant differences from each other – yet we want to refer to the theory of each as Darwinian. One, I suppose, could think of each of these theories as a type, and "Darwinism" as simply the name for the class which includes all of these more restricted types. Darwinian theory, on this construction, would, however, wind up not being a theory at all. It could not be a theory that anyone ever entertained, ever thought or wrote down. Moreover, with each of the members of the class one would have the same problem as with Darwin's initial theory: none remained static over time. Tokens, types, and classes – like the old biological taxa – do not capture the dynamic and developmental character of virtually every scientific theory.

Evolutionary biology itself suggests another, more perspicuous model of theories. We might regard Darwin's theory as an historical individual that changed over time, evolving during its inventor's own life and giving rise to a number of descendent theories – those of Romanes and Haeckel, for instance. The original theory and its offspring – instead of being construed as tokens of the same type – would be reckoned as members of the same species. They would be Darwinian theories in the same way that various individual finches are members of the same species – or, if different enough, members of the same genus, from which they have split off. This might seem a small difference from the type-token approach, but it certainly allows for change of theory over time, without that theory losing its historical identity. Further, it suggests an important analogy that is quite useful in thinking about the fate of theories, this being the genotype-phenotype distinction. For example: at any one time, the theory that Darwin contemplates (even if in a fragmentary way), describes to a friend, summarizes in an essay, or elaborates in a book – all of these might be understood as expressions of the same underlying theory; they would be incomplete phenotypic expressions of an encompassing genotype. Simultaneously, it would be possible to regard the genotype as evolving. Moreover, the phenotype would have distinctive individualizing characters – a stumbling oral expression, for instance, or descriptions directed to a particular audience, or a book with more comprehensive display of the underlying genotype. In this fashion, one could discriminate the basic theory from the momentary embodiment – but, again, without supposing the basic theory to be frozen in some abstract space.

This historical model would have other interesting consequences for dealing with theories. We would, with better conscience, be able to undertake a regressive historical analysis – beginning with present neo-Darwinian theory. We might trace back in a diachronic way the developmental course of a theory-lineage yet with a built-in safeguard against assuming that the nether ends of the lineage must have the same morphological form. Indeed, the beginning and end points might look rather different, but we would still be justified in speaking of present theory as, at least, neo-Darwinian: the justification here being a causally productive one, not one of strict resemblance. We would also be encouraged to conduct a synchronic analysis, an historical investigation of the original intellectual and psychological environment of this species-like theory, of the kinds of pressures it was under, and of the causes that gave the theory its particular shape. This would allow the historian to better understand why a theory had the features it did. Abstract entities, after all, are not subject to alteration by the rough and messy environment of the real world; but historical theories are. Other advantages accrue to an historical model, but I've discussed them elsewhere.[60]

Darwin's theory of evolution by natural selection gave rise to contemporary theories of species change, with their reliance on sophisticated assumptions about population genetics, ecological dynamics, and levels of selection. The trajectory of Darwin's theory, sighted back from our vantage, appears to be a straight line shot from 1859 to the present. And in reading the *Origin of Species* today, it is not hard for biologists to recognize a kindred spirit. Just as Darwin himself was astounded about how much "old Aristotle" knew of biology – when he was presented with a new translation of the *Parts of Animals*[61] – so modern theorists admire the theoretical ingenuity, systematic understanding, organic awareness, and ecological sensitivity demonstrated by old Darwin himself. Darwin's modernity can be understood without assuming perfect identity with contemporary theory, but only if we assume the historical model of theories. If we took seriously the idea that theories were comparable to organic beings, then we might come to appreciate even more one of the concluding observations of the *Origin of Species:*

When we no longer look at an organic being as a savage looks at a ship, as at something wholly beyond his comprehension; when we regard every production of nature as one which has had a history . . . how far more interesting, I speak from experience, will the study of natural history become![62]

If we substitute "history of science" for "natural history," then Darwin's considerations will, I believe, hold good as well for the more modest study conducted in this chapter.

NOTES

1. Imre Lakatos, "History of Science and Its Rational Reconstruction," in *Philosophical Papers Volume 1: The Methodology of Scientific Research Programmes* (Cambridge: Cambridge University Press, 1978), p. 119.
2. Popper distinguished his position from Plato's by holding that such abstract entities as theories were, nonetheless, products of the human

mind. I do not think this emendation alters the liabilities of the historiographic assumption that it supports. See Karl Popper, "On the Theory of the Objective Mind," in *Objective Knowledge* (Oxford: Oxford University Press, 1972).

3. Stephen Jay Gould, "Eternal Metaphors of Palaeontology," *Patterns of Evolution as Illustrated in the Fossil Record*, ed. A. Hallan (New York: Elsevier, 1977), p. 13.
4. Stephen Jay Gould, *Ever Since Darwin* (New York: Norton, 1977), p. 45.
5. Peter Bowler, *Theories of Human Evolution* (Baltimore: Johns Hopkins University Press, 1986), p. 41.
6. Charles Darwin, *On the Origin of Species* (London: Murray, 1859), p. 489.
7. Charles Darwin, Notebook E, MS p. 95, in *Charles Darwin's Notebooks, 1836–44*, ed. Paul Barett, Peter Gautrey, Sandra Herbert, David Kohn, and Sydney Smith (Ithaca: Cornell University Press, 1987), pp. 422–23.
8. Darwin, *Origin of Species*, p. 106.
9. Ibid., p. 337.
10. Ibid.
11. Charles Darwin, Notebook B, MS p. 74, in *Charles Darwin's Notebooks*, p. 189.
12. Charles Darwin to George Waterhouse (July 26 and 31, 1843), in *Correspondence of Charles Darwin*, ed. Frederick Burkhardt and Sidney Smith, 10 vols. to date (Cambridge: Cambridge University Press, 1985–), vol. 2: 375–78.
13. See, for example, Darwin's discussion (*Origin*, p. 204) of the mammalian eye as an organ of great complexity and thus perfection.
14. Ibid., pp. 115–16.
15. Charles Darwin, *A Monograph of the Sub-Class Cirripedia: The Balanidae* (London: Ray Society, 1854), p. 20.
16. Charles Darwin, *The Origin of Species by Charles Darwin: A Variorum Text*, ed. Morse Peckham (Philadelphia: University of Pennsylvania Press, 1959), p. 221.
17. Charles Darwin, loose note, DAR 505.9 (2), MS 200, held in the Papers of Charles Darwin, Cambridge University Library.
18. Darwin, *The Origin of Species: A Variorum Text*, p. 221.
19. See my discussion of this problem in *Darwin and the Emergence of Evolutionary Theories of Mind and Behavior* (Chicago: University of Chicago Press, 1987), pp. 172–76.
20. See, for example, Robert J. Richards, *The Meaning of Evolution: The Morphological Construction and Ideological Reconstruction of Darwin's Theory* (Chicago: University of Chicago Press, 1992), pp. 84–90; and "The Moral Foundations of the Idea of Evolutionary Progress: Darwin, Spencer, and the Neo-Darwinians," *Evolutionary Progress*, ed. Matthew Nitecki (Chicago: University of Chicago Press, 1988). Dov Ospovat, in a book of singular perception, had sounded the first warning that the neo-

Darwinian reading of Darwin recast him in a modern light that kept in shadow the progressivism of his theory. See Dov Ospovat, *The Development of Darwin's Theory: Natural History, Natural Theology, and Natural Selection, 1838–1859* (Cambridge: Cambridge University Press, 1981).

21. See Stephen Jay Gould, *Wonderful Life: The Burgess Shale and the Nature of History* (New York: Norton, 1989), pp. 257–58: "The logic of the theory pulled in one direction, social preconceptions in the other. Darwin felt allegiance to both and never resolved this dilemma into personal consistency."

22. Michael Ruse, *Monad to Man: The Concept of Progress in Evolutionary Biology* (Cambridge: Harvard University Press, 1996), p. 160.

23. Ibid., p. 165.

24. Ibid., p. 167.

25. Ibid., p. 166.

26. Michael Ghiselin, "Review of *The Meaning of Evolution*," *Systematic Biology* 41 (1992): 497–99. See also G. Nelson, "Review of *The Meaning of Evolution*," *Systematic Biology* 41 (1992): 496–97.

27. Erik Nordenskiöld, in the first decades of the twentieth century, judged Haeckel's *Naturliche Schöpfungsgeschichte* (1868), which went through twelve German editions and countless translations, "the chief source of the world's knowledge of Darwinism." See Erik Nordenskiöld, *The History of Biology* (New York: Tudor Publishing, [1920–1924] 1936), p. 515.

28. See the introduction to a modern edition of Haeckel's *Die Welträtsel*, ed. Olof Klohr (Berlin: Akademie Verlag, 1961), pp. vii–viii.

29. See the introduction to *The Origin of Species by Charles Darwin: A Variorum Text*, ed. Morse Peckham (Philadelphia: University of Pennsylvania Press, 1959), p. 24.

30. See Haeckel's original expression of this principle in Ernst Haeckel, *Generelle Morphologie der Organismen*, 2 vols. (Berlin: Georg Reimer, 1866), vol. 1, p. 265.

31. Haeckel's own estimate of evolutionary progress does not differ from Darwin's own. See ibid., pp. 257–66.

32. See Stephen Jay Gould, *Ontogeny and Phylogeny* (Cambridge: Harvard University Press, 1977), p. 70: "Darwin had accepted the observations of von Baer – a flat denial of recapitulation."

33. Peter Bowler, *The Non-Darwinian Revolution: Reinterpreting a Historical Myth* (Baltimore: Johns Hopkins University Press, 1988), p. 198.

34. Gould, *Ontogeny and Phylogeny*, pp. 77–78: "His [Haeckel's] evolutionary racism; his call to the German people for racial purity and unflinching devotion to a 'just' state; his belief that harsh, inexorable laws of evolution ruled human civilization and nature alike, conferring upon favored races the right to dominate others; the irrational mysticism that had always stood in strange communion with his brave words about objective science – all contributed to the rise of Nazism."

35. Charles Darwin to Asa Gray (September 10, 1860), *Correspondence of Charles Darwin*, vol. 8, p. 350.
36. John Lubbock to Francis Darwin (January 2, 1896), ibid., p. 280n.
37. Friedrich Tiedermann, *Zoologie, zu seinen Vorlesungen Entworfen*, 3 vols. (Landshut: Weber, 1808–1814), vol. 1, p. 65.
38. Richard Owen, *Richard Owen's Hunterian Lectures, May–June 1837*, ed. Philip Sloan (Chicago: University of Chicago Press, 1992), MS 98–99, p. 192.
39. Darwin, Red Notebook, in *Charles Darwin's Notebooks*, MS 129–33, pp. 62–63. Darwin carried the idea that the species was comparable to an individual with a definite life span into the first part of his Notebook B. See Darwin, Notebook B, in *Charles Darwin's Notebooks*, MS 22, p. 177.
40. Ibid., MS 1, p. 170. The wedge brackets indicate words that Darwin deleted.
41. Ibid., MS 78, p. 190.
42. Darwin, *Origin of Species*, p. 444. Darwin, nonetheless, provided what he thought convincing evidence for this proposition. See ibid.
43. Ibid.
44. Ibid., pp. 444–47.
45. Charles Darwin, "Essay of 1844," DAR 7, MS 163b–64. Wedge brackets indicate words that Darwin deleted. This manuscript is held in the manuscript room of Cambridge University Library. It differs, though slightly, from the version published by Darwin's son. See Charles Darwin, *The Foundations of the Origin of Species: Two Essays Written in 1842 and 1844 by Charles Darwin*, ed. Francis Darwin (Cambridge: Cambridge University Press, 1809), p. 230
46. Ghiselin, "Review of *The Meaning of Evolution*," p. 499.
47. Darwin, *Origin of Species*, p. 338.
48. Darwin, *Origin of Species . . . a Variorum Text*, p. 704.
49. Ibid., p. 345.
50. Darwin, *Origin of Species*, p. 345.
51. Ghiselin, "Review of *The Meaning of Evolution*," p. 498.
52. Early and late, it is quite patent that Darwin believed the embryo recapitulated the *adult* stages of its ancestors. The texts to notes 45, 47 48, and 49 make that conviction explicit. Moreover, his two principles of heredity, discussed earlier, have as an intended logical consequence that the embryo passes through the adult forms of its predecessors.
53. Darwin, *Origin of Species*, p. 444.
54. Ibid., p. 443.
55. Ibid., p. 447: "Whatever influence long-continued exercise or use on the one hand, and disuse on the other, may have in modifying an organ, such influence will mainly affect the mature animal, which has come to its full powers of activity and has to gain its own living; and the effects thus produced will be inherited at a corresponding mature age. Whereas

the young will remain unmodified, or be modified in a lesser degree, by the effects of use and disuse."

56. Ibid., p. 444.
57. Ibid., p. 338.
58. See the letter to Gray quoted earlier, note 33.
59. Dov Ospovat quite convincingly argued that through 1844, Darwin believed that evolution produced perfect adaptations. See *The Development of Darwin's Theory*, pp. 60–86.
60. See Robert J. Richards, *Darwin and the Emergence of Evolutionary Theories of Mind and Behavior* (Chicago: University of Chicago Press, 1987), pp. 579–93.
61. See Darwin's letter to Aristotle's translator, William Ogle (February 22, 1882), in *Life and Letters of Charles Darwin*, ed. Francis Darwin, 2 vols. (New York: D. Appleton and Co., 1891), vol. 1, p. 427.
62. Ibid., p. 485–86.

Part Two

Laboratory and Experimental Research

The Nature and Use of Evidence

Chapter 5

Down the Primrose Path

Competing Epistemologies in Early Twentieth-Century Biology

DAVID MAGNUS

INTRODUCTION[1]

When most people hear the expression "natural history," they imagine a sort of amateur bird-watcher or butterfly collector who runs around capturing, describing, and identifying species – in very definite contrast to our image of a proper "biologist" or "scientists." Journalist-historian Lynn Barber has expressed this most clearly. "By the end of the [nineteenth] century, all professionals were calling themselves 'biologists', and 'naturalist' was becoming the hallmark of the sort of sentimental amateur who referred to birds as his feathered friends."[2] To see natural history as something less than science is very common among biologists, philosophers, and historians of science.

Accounts by historians of biology working on the period from 1890 to 1910, a crucial time for understanding biology, embody many of the judgments made by philosophers and biologists about natural history.[3] In particular, a picture of natural history and the development of modern biology out of a split biological community has gained some prominence in the history of biology. I shall refer to this as "the standard picture." This picture or view holds that a modern biology which is (a) experimental, (b) quantitative, (c) more narrowly focused in topic, and (d) more concerned with micro-mechanisms and with physical mechanisms generally, emerged when a group of young biologists rejected the older, less successful tradition of natural history, which was (a) descriptive, (b) qualitative, (c) speculative, and (d) concerned solely with grand evolutionary or taxonomic questions. The failure of the methods of natural history to suc-

91

cessfully address the significant biological problems of the day is cited as the reason for the transformation.

I argue that in at least one important context, when one can make a distinction between a group identified as natural historians and a more experimental group, the differences between them should be understood as reflecting differences in epistemologies. Different groups are committed to different scientific virtues, which are seen as essential to sound research. Contrary to the characterizations of the standard picture, the epistemology of the naturalists was not purely qualitative, nor purely descriptive, and most importantly it was not unjustifiably speculative. Contrary to the view that natural history was unable to deal with significant biological problems, it was actually better able to solve problems, in the case study I consider, than its competitor, an epistemology which valued experimental control over all else.[4]

The epistemology of the natural historians stressed the importance of a *variety* of lines of evidence for arriving at conclusions. This scientific virtue is sometimes identified as the consilience of inductions (hereinafter consilience).

The concept of consilience of inductions was developed by the nineteenth century historian and philosopher of science William Whewell. Whewell was concerned with establishing principles of inductive reasoning which fit with the historical practice of scientists. In place of traditional Baconian induction, in which the only "theoretical" activity involved self-evident generalizations from immediate experience, Whewell suggested that the best science involved the making and testing of bold hypotheses. For Whewell, the best support for a theory is when a number of different kinds of evidence all support it (or confirm it).

> We have here spoken of predictions of facts *of the same kind as* those from which our rule was collected. But the evidence in favour of our induction is of a much higher and more forcible character when it enables us to explain and determine cases of a *kind different* from those which were contemplated in the formation of our hypothesis. The instances in which this has occurred, indeed, impress us with a conviction that the truth of our hypothesis is certain . . .
>
> Accordingly the cases in which inductions from classes of facts altogether different have thus *jumped together*, belong only to the best established theories which the history of science contains. And as I shall have occasion to

refer to this peculiar feature in their evidence, I will take the liberty of describing it by a particular phrase; and will term it the *Consilience of Inductions.*[5]

Whewell's views on method were subsequently overshadowed by those of Mill. But Whewell's work was influential in certain circles, and had an effect on Charles Darwin. Michael Ruse and Paul Thagard have argued that Whewell's views – especially the concept of consilience – played an important role in the discovery and justification of Darwin's theory.[6] Regardless of the causes, Darwin came to adopt a very similar view of method, structuring his argument in such a way as to generate a consilience of inductions.

Another virtue associated with consilience is breadth. There is a tight connection, conceptually, between consilience and breadth. Consilience can be seen as a virtue of the evidential basis for a model or theory. Generative consilience is the virtue associated with the evidence that leads to the discovery of a theory. Confirmative consilience is a virtue of the evidence cited as justification for a theory.[7] In contrast, breadth has to do with the scope of a theory or model. Some theories are capable of relating a diverse number of areas, or lines of evidence, while some are restricted to answering questions in a narrower range of problems or addressing a smaller number of lines of evidence. After Darwin, it became accepted among most naturalists that the scope of evolutionary theories needed to be fairly broad, capable of explaining and relating a number of diverse areas.

Following Darwin, naturalists came to see consilience and breadth as the key scientific virtues. They often identified this as "the Darwinian method." Another virtue, which Herschel (and Whewell) identified and which seems to have become part of the Darwinian method, was "conversion."[8] It was seen as very strong support for a theory when problematic data could be predicted or expected as an unintended consequence of a theory. For example, Darwin thought that the embryological fact that most early vertebrate embryos are almost indistinguishable would be a problem for any evolutionary theory (since evolution is concerned with "permanent" features of organisms). However, Darwin realized that natural selection offered an explanation of this embryological fact. Since embryos all face roughly the same "conditions of life," we do not expect much differentiation. But, as adult forms face greatly different environments, selection leads to differentiation.

At the same time, in place of the naturalist epistemology and its focus on a variety of lines of evidence and a multiplicity of techniques, a number of biologists advocated the centrality of an alternative set of scientific virtues. One of the most important, according to this more experimental epistemology, is *experimental control.* These biologists felt that biological progress was hampered by a lack of rigor, that the characteristics that had made chemistry and physics so successful were often lacking. Many of the lines of evidence that were seen as important to natural historians necessitated a central role for individual judgment, which others saw as too subjective and speculative to be allowed any part in a full-blown scientific enterprise. According to this alternative view, legitimate evidence lacked this subjective element; hence it was claimed that biology would be put on a more objective footing by employing replicable, experimental techniques. In this way, all knowledge could be placed on a secure basis, since it would rest on evidence open to anyone to check.[9]

Although there were attempts to maintain an emphasis on both virtues – consilience and control – they tended to pull in opposite directions, and the tension was particularly acute in evolutionary biology. There was no way to reconcile most of the traditional methods of natural history with the new epistemological requirements. Because of this, one had to either give up replication, or abandon most of the sources of evidence for evolutionary conclusions and hence abandon consilience. Advocates of the experimental epistemology took the leap and abandoned all nonexperimental forms of evidence.

Because the number of lines of evidence to be taken into account must be limited, on their view, the scope of biological theories was similarly restricted.

Instead of theories which can explain all conceivable phenomena of development (as Weismann and Roux wanted) or as many empirically observed phenomena as possible (as Hertwig allowed) of the available phenomena, the Americans sought to accumulate as many definite facts as possible.[10]

There is much less concern with generating large scale theories to 'explain' or to 'give the causes of' all of living phenomena at once.[11]

Thus, the scope of theories and arguments was greatly narrowed – the breadth which was seen as a virtue by the naturalists was impossible within the experimentalist epistemology.

Another virtue associated with the experimental epistemological package is the virtue of parsimony. According to this view, the goal of biology is ultimately to explain everything in terms of the micro-mechanisms of heredity and variation – whatever they are. Hence, genetics represents the central domain of biology, and it should be explored in an experimental and quantitative fashion. Ultimately, all biological phenomena should be explained in terms of knowledge of genetics.

These two epistemological views are reflected in an important debate which took place in the early twentieth century. The origin of diversity of species had been an acute problem in the nineteenth century, one that seemed to reach a resolution with the development of Darwin's theory. However, even in the early days of the Darwinian reign there were controversies over the exact mechanisms responsible for the production of species. One controversy centered on whether geographical isolation was necessary for evolution. Darwin had changed his views, initially being sympathetic to seeing isolation as central, but eventually denying it much of a role.[12] A controversy among the next generation of Darwinians ensued. This controversy was fading by the end of the nineteenth century as an increasingly large number of naturalists (including Darwinians) came to accept the view that isolation in some form was essential to speciation.[13] However, a new biological view, Hugo de Vries's Mutation Theory, gave impetus to increasing debate and called the collective judgments of naturalists into doubt.

THE MUTATION THEORY

In 1901 and 1903 de Vries (one of the co-rediscoverers of Mendel's Laws) published his mutation theory, which was the culmination of his experiments on *Oenothera*. In these experiments de Vries found that two forms apparently derived from *Oenothera Lamarckiana* both bred true. These he identified as *O. brevistylis* and *O. laevifolia*. Each of these was quite distinct from *O. Lamarckiana*. *O. brevistylis* was short-styled, while *O. laevifolia* had smoother leaves and "prettier foliage" than the parent plant.[14]

When he bred *O. Lamarckiana*, de Vries was able to produce many more plants which were morphologically distinct from the parent stock and which bred true when separately raised. Over time, many new forms were developed and isolated. De Vries considered each of these forms to be genuine species.

De Vries therefore felt that he had been witness to the production of new species – and a large number of them. *O. lamarckiana* seemed capable of "throwing off" a great number of new species. De Vries was well aware that this was unusual. None of the other plants he bred showed any tendency to produce new species in the manner of *Oenothera*. He put together these findings into the mutation theory.

Most Darwinians held that the relevant variations which natural selection acted upon were small differences. However, a different view had arisen almost as soon as Darwin had published his theory. Thomas H. Huxley thought it likely that the "monsters" or "sports" which were known to occur sometimes, could be the variations which natural selection acted upon. Although most Darwinians did not accept this view, Francis Galton's work developing the notion of regression led him to adopt a similar position. He argued that only discontinuous variations could lead to evolution by avoiding regression toward the mean.[15] One of the most forceful proponents of the view that variations (or at least the relevant ones) were discontinuous was William Bateson. After engaging in traditional morphological work on *Balanoglossus* for a time, Bateson turned to the study of variation. Bateson thought that the small "fluctuating" or "continuous" variations Darwinians posited would be inadequate as a basis for natural selection. He argued that discontinuous variations were the common ones responsible for evolutionary change, while continuous or minute variations were irrelevant to evolutionary questions.[16] Bateson went on to become a leading proponent of Mendelism in England.

De Vries interpreted his work on *Oenothera* as indicating that the Darwinians were wrong, and that Bateson was correct. Discontinuous variations were the relevant evolutionary material. Natural selection would still play a minor role, eliminating unfit mutations; but the cause of organic transformation should be understood as mutation. De Vries, like Bateson, saw the fact that inheritance was particulate as further support for the theory.

For de Vries, new species occurred in large, discontinuous jumps, taking place among several individuals at once. New species were produced alongside the parent stock. There was no role for geographical isolation, according to this view. Evolution was always sympatric.

De Vries's theory was warmly received by much of the biological community, particularly among biologists engaged in experimental studies of variation, such as Charles Benedict Davenport, Daniel Trembly MacDougal, Reginald Ruggles Gates, T. H. Morgan, and Bateson. As Allen has argued, de Vries's methods were at least partly responsible for the warm reception he received. De Vries suggested that experimental methods should be the starting point for investigating the laws governing evolution. He thought that traditional "comparative" methods had been useful in the past for establishing the *fact* of evolution, but to understand the actual mechanisms underlying it required experimental data. In *The Mutation Theory* he suggested that *after* a theory was constructed and supported on the basis of experimental data, other kinds of evidence could be considered to see if they fit with the theory. There was no discussion of what would happen if there were a lack of fit between the kinds of data, but de Vries's practice was suggestive. The facts of paleontology were seen to accord well with the mutation theory, and so he accepted these results. Since the mutation theory implied that isolation was irrelevant to evolution, the theory was inconsistent with the results of biogeography (which offered facts that could only be accounted for by accepting the role of isolation). De Vries ignored these arguments.

The origin of species has so far been the object of comparative study only. It is generally believed that this highly important phenomenon does not lend itself to direct observation, and, much less, to experimental investigation.

The object of the present book is to show that species arise by saltations and that the individual saltations are occurrences which can be observed like any other physiological process . . .

In this way we may hope to realize the possibility of elucidating, by experiment, the laws to which the origin of new species conform. The results of these studies can then be compared with those which have been obtained with systematic, biological and particularly with paleontological data.[17]

In *Species and Varieties* (1904) de Vries told the history of biology as a progression from Darwin, who had some evidence for evolution but had to rely on insufficient breeding evidence for his theory, to the "present," which provides ample evidence. The progression also followed from older methods to newer, better methods.[18]

As the discussion of de Vries's views about species suggests, he thought experimental data were necessary for determining vital facts about the nature of forms. Because of his reliance on experimentation, and the role it played as the principal method of study, many biologists accepted the truth of de Vries's theory. As Davenport claimed in a review, "The great service of de Vries's work is that, being founded on experimentation, it challenges to experimentation as the only judge of its merits."[19]

JORDAN'S 1905 PAPER

In 1905 David Starr Jordan, the president of Stanford University, a prominent naturalist, and perhaps America's leading ichthyologist, published his article "The Origin of Species through Isolation" in *Science*. In this article he tried to show, contra de Vries, that isolation was essential to speciation.

Jordan began his article by considering the origin of species in a given location. This very naturally drove him to recognize a curious fact. In regions broken up by barriers like mountains, or where streams are isolated, there is an increase in the amount of speciation, although the range of each species will be quite small. By contrast, the number of genera or families will be smaller in such regions, since free migration is prevented. Jordan noticed that in barrier-free regions, there is a greater variety of families but less speciation, as the range of a species is greatly increased. These puzzling phenomena needed to be explained, and could be accounted for most directly by recognizing the role that isolation plays in speciation.

Jordan also noted a regularity in the geographical distribution of most species, a regularity associated with the range of each species. One virtually never finds closely related species in the same streams, but only in neighboring, isolated streams. This led Jordan to conclude that he had found something which he thought should be "raised to the dignity of a general law of distribution." This would become known as Jordan's Law:

> Given any species in any region, the nearest related species is not likely to be found in the same region nor in a remote region, but in a neighboring district separated from the first by a barrier of some sort.[20]

In establishing the scope and lawlike nature of this regularity, Jordan cited numerous instances of its occurrence in the distribution of fish in the U.S., Hawaii, and the South Seas. For Jordan, it was important that theoretical views be supported by different lines of evidence. Therefore, Jordan cited a "circular letter" he had sent to some leading ornithologists, to ascertain whether he would find further support.[21] He published the letter along with certain "typical responses." The letter read:

> In considering the proposition that species in general arise in connection with geographic or topographic isolation, will you kindly answer briefly the following questions?
> 1. Do two or more well founded subspecies ever inhabit (breed in) the same region? If so, give examples.
> 2. If so, how do you explain the fact?
> 3. Would you regard a form as a 'subspecies' if coextensive in range with the species with which it intergrades?[22]
> 4. Are there cases where two species inhabiting exactly the same region are closely related, and more closely than any other species is to either one? If so, give examples.[23]

Jordan sent this letter to ten people. In addition to the seven responses he published, he sent letters to Joel A. Allen, Robert Ridgeway, and Theodore S. Palmer.[24]

So, in addition to his ichthyological data, Jordan had the authority of many of the leading zoologists of his day, providing further examples of his law of distribution. He had additional examples drawn from ornithology and mammology to bolster his own work on the distribution of fishes. His circular letter provided important support for his views. This technique was familiar to Jordan from his work with the Fish Commission. Further, his circular letter and the answers he received seemed to indicate no compelling counterinstances to his law. This was partly due to several explanatory devices available to the "isolationists." One was to attribute problem cases to a lack of detailed investigation on the part of naturalists (Stejneger as well as Jordan suggested this)[25]. Another explanation was in terms of "seasonal isolation," as species which are closely related inhabit the same region but breed at different seasons (Stejneger, Fisher). Jordan further cited the possibility of reinvasion after the barriers have been removed, which would give the appearance of two closely related species arising in the same region, but would really reflect the migration of one species into the domain of another. Of course, all of the examples and individuals Jordan cited were from zoology, and the problem cases from botany had not yet been considered. With all of this evidence, and the broadness of its application, Jordan felt confident that he had given his observed regularity in nature the status of a general law of distribution.

This law of distribution, plus the fact of increased speciation in barrier-laden regions, were the two chief facts which Jordan thought required recognizing the role of isolation in speciation. The alternative mutation theory, which had a new species arise alongside its parent, simply did not reflect the evidence of the distribution of species to be found in nature. If the mutation theory were true, one would expect the most closely related species to arise in the same region as its parent, rather than in a neighboring region; and one would expect that since barriers or isolation were irrelevant to speciation, there would be less speciation in regions broken up by barriers, just as there were fewer genera or families.

In addition, Jordan's theory explained the fact that most characteristics which distinguish a species seemed to serve no current adaptive function. Jordan recognized this, citing Kellogg's statistical work to show that the distinguishing traits in honeybees and beetles were nonadaptive.[26] This was an important problem for a Darwinian like Jordan, and one of the virtues of the mutation theory was that it could account for this problem. Since Jordan thought natural selection was not operative in many cases of speciation, as isolated groups diverge from one another, he could also solve the problem. One would actually expect to find the slight nonadaptive morphological differences one does find, according to this view.

At this point Jordan had no clear mechanism for the divergence of characteristics once two groups were isolated. He seemed to think that the difference might be due to slight differences in the parent stock, or to differences directly resulting from the environment.[27] That is, variation might be partly caused by the environment, and since no two regions are exactly alike, the variation in the isolated groups would be different, and hence result in divergence and eventually speciation. A third possibility involved mutations. In an interbreeding population, a "sport" would be lost in the population. But if the sport (and perhaps a few like it) were isolated, its characteristics could become the dominant ones in the isolated group.

Whichever of the three mechanisms was actually operable (or some combination, plus sexual selection) it would lead to speciation with no differences in adaptive value. Of course, if there were sufficiently large environmental differences, Jordan thought natural selection would produce adaptive differences in the forms of the isolated species.

Jordan thus presented a powerful case for the role of isolation in speciation, and its importance to Darwinian theory. More generally, he showed that attention to the actual distribution of species could and did shed light on the origins of speciation and its causes, and that the "semi-contemptuous attitude of morphologists and physiologists towards species mongers and towards out-door students of nature generally" was not justified.[28]

In addition to the influence of Darwin on the content of Jordan's theory, the structure of his argument also had a Darwinian character. The overall structure of the argument was to begin with a variety of facts that needed to be explained, then to show how these could all be explained by appealing to the role of isolation. Like Darwin's *Origin*, Jordan's article included a consideration of counterexamples. He appealed to the virtue of conversion just as Darwin did. In Jordan's argument, the existence of nonadaptive traits was converted into an expectation of his (broadly Darwinian) view, rather than a problem.

The debate about the role of isolation and the mutation theory was given an additional impetus by Clinton Hart Merriam's address to the zoological section of the AAAS. This was published in *Science* in 1906 under the title, "Is Mutation a Factor in the Evolution of the Higher Vertebrates?" Where Jordan had focused on developing arguments in favor of his view of evolution, Merriam rose to attack the mutation theory.

Speaking of the mutation theory, Merriam asked, "Are we, because of the discovery of a case in which a species appears to have arisen in a slightly different way – for after all the difference is only one of degree – to lose faith in the stability of knowledge and rush panic stricken into the sea of unbelief, unmindful of the cumulative observations and conclusions of zoologists and botanists?"[29]

Merriam thought not – that de Vries's experiments did not indicate anything very general about evolution.

> To my mind the striking fallacy of de Vries's argument is his assumption that because it can be shown that sports occur in nature, and that in rare cases species arise there-from, then the theory that species are produced by the progressive development of slight individual variations must be abandoned.[30]

Merriam turned from his own views to the inadequacy of the mutation theory, and utilizing Jordan's work, undertook a detailed discussion of isolation. In discussing the origins of subspecies, he wrote, ". . . subspecies in nature do not occupy the same ground with the parent form, but an adjacent area; hence, it is hard to see how they could fulfil [sic] his [de Vries's] geographic requirement, which is that forms arising by mutation occur side by side with the original stock."[31]

Merriam argued that the only way to understand the production of subspecies was to study their geographical distribution. Failure to do this, or to study a sufficiently large area (so that it includes transitions from one fauna to another) accounted for most misconceptions concerning the nature of subspecies. Since Jordan had shown how important isolation was to the production of species, de Vries's theory would fail in most cases, according to Merriam.

Merriam believed this case to be typical of the kinds of puzzles a naturalist encounters in trying to understand the details of particular lineages. In these cases, "without a comprehensive knowledge of the relationships and geographic distribution of the group as a whole and of its component species and subspecies, there is little hope of arriving at correct conclusions."[32]

In championing the importance of knowledge of geographical distribution, Merriam had a broad concept in mind, following Jordan.

The term geographic distribution must not be taken to mean merely the area a species occupies, to be shown by a color patch on a map, but includes a comprehensive knowledge of the geographic environment, taking into account the climate and the aspects of nature with which each species is associated and by which it is profoundly impressed. Moritz Wagner, in a paragraph recently quoted by Doctor Jordan, said: It is 'the study of all the important phenomena embraced in the geography of animals and plants, which is the surest guide to the study of the real phases in the processes of the formation of species.'[33]

RESPONSES

Jordan's article constituted an attack on the mutation theory. If Jordan was correct, then the mutation theory had to be wrong because it did not allow any role for geographical isolation in evolution. The facts of distribution cited by Jordan had no explanation on the de Vriesian account. C. H. Merriam utilized these arguments along with other biogeographical evidence to forge a direct attack on the mutation theory, and on de Vries's work in particular. This led to a clash between mutation theorists and isolationists. The predominant response was not to attack the facts cited by Jordan and Merriam but to question their entire methodology. The practice of natural history was claimed to be speculative, subjective, and nonscientific. Because of this, there was no need to give a detailed refutation of the evidence and arguments marshaled in favor of the naturalists' views. Instead, only certain evidence met the standards genuine science demanded; and this (experimental) evidence pointed toward the mutation theory. The isolationists had to defend the way they did science, to present their views of what constituted proper methods. At a time when biologists (including natural historians) were debating everything from what constituted a problem and how it should be solved, to details of morphology and cell structure, the question of the proper method for solving evolutionary problems took on added significance. Following Merriam's attack, debate continued in the pages of *Science* over the validity of the mutation theory. I will focus primarily on issues surrounding the role of isolation, and on epistemological arguments that undercut the arguments of the isolationists.

Shortly after Merriam's address, Thomas Wayland Vaughan delivered an address at the Biological Society of Washington which was subsequently published in *Science*. Vaughan, who had received his Ph.D. from Harvard, was a geologist by training with a strong interest in the physical sciences. At the time of his address, he was with the U.S. Geological Survey. In his article, Vaughan contrasted what he took to be the three rival accounts of the origin of species: the mutation theory, the Darwinian theory, and the neo-Lamarckian theory. In sketching out the mutation theory, Vaughan argued that the problem with Merriam's arguments was that they were necessarily inconclusive: the facts of distribution are consistent with any theory, so only by experimentation could one decide among the alternative theories. "His facts seem just as plausibly explicable on the basis of the Darwinian hypothesis or that of de Vries."[34] Again, this was Vaughan's general view about this style of argument – showing that "a body of facts were in harmony with the *assumption'* " does not get one very far.[35] He approvingly quoted from de Vries, "it is absolutely impossible to reach definite conclusions on purely morphological evidence. This is well illustrated by the numerous discords of opinion of different authors on the systematic worth of many forms."[36] "Until various physiological tests of the kind referred to by de Vries have been made, more than an hypothetical explanation of the facts presented by Dr. Merriam is impossible."[37] (By implication, this would apply to Jordan's arguments as well.) So Vaughan attacked Merriam's arguments on the ground that the kind of evidence he took into account was always open to interpretation, and hence shed no real light on which theory is correct. Only experimental evidence, preferably of a physiological nature, would be sufficient. Vaughan treated the issue of isolation in a similar manner:

> I think that it is not necessary to give a special discussion of isolation as a factor in evolution, as it does not affect the validity or invalidity of any one of the hypotheses stated. Isolation is a passive, not an active factor; its importance, however, is beyond question.[38]

Charles Stuart Gager was a professor at the New York State Normal College. He had received his Ph.D. from Cornell, and served as editor for several prominent botanical journals as well as the *Journal of Applied Microscopy*. He responded to the attacks launched by Merriam and other naturalists who participated in the debate, particularly Arnold E. Ortmann. For Gager, the mutation theory of de Vries had the most secure foundation possible, namely an experimental one. Merriam had claimed on the basis of "extended field studies" that minute variations gradually produce new forms. Gager pointed out that de Vries "shows" that

'long-continued selection, alone, has absolutely no appreciable effect' in changing the inner nature of a species or of a race, whereas there is experimental evidence of another factor by means of which such a change is accomplished.[39]

For Gager, the only way of making scientific progress was through the experimental method. Because experiments can be repeated by anyone, they lead to an objectivity and certainty that other methods (the methods of Ortmann, Merriam, and Jordan) do not have. "The beauty of experiment is that it convinces *all*, because given results may be produced by all alike at will."[40] The other advantage of de Vries's experiments was that they allowed for the direct detection of the processes of evolution.

The beauty of de Vries's method is that it is possible 'to arrange things so as to be present when nature produces . . . these rare changes.' In this way it is *known* (not a matter of opinion) that 'the mutation took place at once . . . No intermediate steps were observed . . . Not a single flower on the mutated plant reverted to the previous type.[41]

Merriam's field experience suggested that gradual variation was involved in the formation of different groups, as he found a number of species (or subspecies) to intergrade. Gager quoted de Vries, who pointed out that these cases may be ones in which there is apparent, but not real, overlap.

'Transitions are wholly wanting, although fallaciously apparent in some instances,' as they appeared to Merriam, 'owing to the wide range of fluctuating variability of the forms concerned, or the occurrence of hybrids and subvarieties.'[42]

Because of this, Merriam's argument was inconclusive. Gager went on to say that only by experimental work on these *apparently* intergrading forms, could information about the actual processes be uncovered. The question was whether or not the possible intergradations were all stages passed through by an evolving organism. De Vries had experimentally established that this was not the case, according to Gager.

> One of the greatest values of de Vries' work was in the fact that he was present when the transition took place, and gives, not a theory at all, but the record of a fact observed again and again.[43]

Gager went on to praise de Vries for introducing a new set of experiments and using that as the sole method of conducting research.

> The mutation 'theory' is still largely a working hypothesis. It is founded almost entirely upon experiment, and can be verified only by the same means.[44]

To Gager, experiments allowed replication by anyone and hence yielded matters of fact, whereas traditional field methods, based upon inductive inferences, were merely "opinion." "For mere opinion and inference, and *a priori* impressions and prejudice, and induction, from field studies and comparative morphology there is absolutely no place."[45]

The arguments offered by Jordan, Merriam, and Ortmann about the role of isolation were given almost no consideration by Gager, since they were based upon fieldwork and systematics. In the end, he contrasted the neo-Darwinian picture of Jordan, Ortmann, and Merriam with the mutation theory.

> How do species originate? A mass of facts suggests that the method is by the natural selection of fluctuating variations, combined with geographical isolation, influence of environment and other factors. But, after all has been written, the undeniable fact remains that no one has yet ever actually observed the origin of a single species in this way. On the other hand, the fact is just as undeniable that a definite and clearly defined type of variation has been actually observed, not once, but often and by many, to arise by a process, equally well-defined and definite, and known as 'mutation.' The case seems perfectly plain that the burden of proof rests with the adherents of fluctuation.[46]

Another attack on Merriam was launched by the famous experimental evolutionary biologist C. B. Davenport. Davenport had received his Ph.D. from Harvard in 1892. As director of the Department of Experimental Evolution of the Carnegie Institution of Washington at Cold Spring Harbor, Davenport was one of the leading mutation theorists, as well as a leader in the eugenics movement and a strong Mendelian. He claimed that the question of the origin of species should be thought of as the origin of the distinguishing or specific morphological characteristics of organisms.[47] The isolationists (Darwinians and neo-Lamarckians) thought that these characteristics arose gradually through the accumulation of small variations. The mutation theorists claimed that they arose "suddenly and completely" in a single generation. Davenport attacked Merriam's general argument that swamping would reduce the effect of mutations. "His [Merriam's] query . . . would not have been asked if he had bred sports and observed their resistance to blending and reversion."[48]

Davenport agreed with Merriam that the general pattern of change in coloration one finds in a number of species of small mammals and nonmigratory birds in North America was a problem for the mutation theory. Davenport wanted to leave it an open question whether these changes constituted differences in species.[49] However, he was willing to admit the possibility that "geographic variation" might produce different species. In these cases, one did not find genuinely new characteristics, but only modifications of old ones – the changes were quantitative rather than qualitative. Davenport claimed that it was possible that qualitative differences may have evolved in a different way than quantitative ones. Hence, the question of interest to evolutionary biologists was which evolutionary process was more common, and in particular cases, which process was at work. For Davenport, this simply meant finding out whether a characteristic intergraded or blended in breeding experiments. If a trait was Mendelian, then the mutation theory must be the correct account of its origin. If it was blending, then a more traditional Darwinian or Neo-Lamarckian account would be required.

The practical way to get at the true nature of characteristics, whether continuous or discontinuous, is by their behavior in inheritance. If, in cross-breeding, a character tends to blend with the dissimilar character of its consort it must be concluded that the character can be fractionized and that intergrades are possible. If, on the contrary, the characteristic refuses to blend, but comes out of the cross intact, as it went in, the conclusion seems justified that the characteristic is essentially intergral and must have arisen completely formed, and hence discontinuously.[50]

According to Davenport, only through breeding experiments could the correct theory of evolution be uncovered in a given case. In experiments he conducted, Davenport found that "most characteristics, but not all, fail to blend and are strictly alternative in inheritance. I interpret this to mean that the characteristic depends on a certain molecular condition that does not fractionize."[51] Hence, contrary to Merriam's view that mutations or sports rarely caused evolution (and never among animals), Davenport claimed that they were the typical cause of new species.

Davenport's view that breeding experiments were the key to understanding evolution obviously ran counter to the biogeographical orientation of Merriam (and of Jordan, Ortmann, et al.). He attacked Merriam for his methodological approach to evolutionary problems.

> Regrettable is Merriam's innuendo directed toward zoologists who have been trained in analytical methods involving the use of the microscope. I am not convinced that analytical training in the laboratory is a less adequate training for tackling the species problem than setting traps and shooting and skinning animals and birds in the field.[52]

Davenport avoided dealing directly with the issue of the role of isolation, and the problems that Jordan's distributional regularities (cited by Merriam) presented to the mutation theory. By reducing the problem to one which could be determined by simple breeding experiments, Davenport sidestepped considerations of any other kind of evidence.

The only contribution to the debate that occurred outside the pages of *Science* was an article by the prominent mutation theorist D. T. MacDougal.[53] MacDougal was the director of the department of botanical research at the Carnegie Institute. He had received his Ph.D. from Purdue in 1897. He was a vigorous defender of de Vries's theories, and had edited the Berkeley lectures that made up de Vries's popular volume *Species and Varieties*. He gave a lecture at the Marine Biological Laboratory at Woods Hole, Massachusetts, in 1906, later published in *Popular Science Monthly*.[54] In this paper, MacDougal attacked the claims made by Jordan, Merriam, and Ortmann and gave a general account of some of the experimental work in which he was engaged. MacDougal saw great strides being made in the study of the origin of species. Work on the "physical basis" of heredity combined with breeding experiments made it possible to understand the underlying mechanisms of evolution. He contrasted this with the traditional methods, particularly taxonomy, which he thought were inadequate for understanding evolution because of the physiological character of inheritance and variation.[55]

As an explorer, do you wish to ascertain the source, direction, character, rate of flow and confluence of a river across your route? Surely, you do not reasonably attempt it by an examination of a single reach, or from a photograph of a single waterfall. Even so surely you may not gauge the possibilities of development, or estimate the potentiality or method of action of groups of characters, embraced in a hereditary strain, guided by dimly recognized forces for thousands of years, by the measurement of a preserved specimen. Physiological problems demand analytic methods of observation of living material.[56]

In contrast, MacDougal noted the "inadequate" treatment of these issues in the debate discussed here. He found Merriam's arguments entirely unpersuasive. Merriam, who wanted to address the issue of the existence of mutations on the basis of field experience, was simply mistaken about the value of fieldwork for problems like evolution, according to MacDougal.

Dr. Merriam does not find any evidence to support the conclusion that species arise by mutation. It would be a matter of great surprise if he had. It would be as reasonable to have demanded of him the solutions of problems of respiration from his preparations and field notes. Once a mutant has appeared, no evidence of its distribution can be taken to account conclusively for its origin.[57]

His own knowledge of mutations came from the only possible source of "reasonable certainty," breeding experiments, that is, "pedigree-culture."[58] The arguments of "the naturalists" about general issues, such as the role of isolation, tended to be overly speculative, according to MacDougal. This was partly because of the tremendous differences between animals and plants.

[T]he 'naturalists', as some zoologists term themselves, having made the greatest number of essays to offer a universal interpretation of the problems of distribution, are to be credited with the greatest number of defenseless assumptions.[59]

MacDougal considered the discussion of the role of isolation an example of "unedifying results" which failed to consider adequately the differences between animals and plants. MacDougal's objection did not focus on the facts of distribution, but on their interpretation. He considered the arguments of Jordan, Merriam, and Ortmann to be a derivation from the *factors* they thought were involved, rather than from geographical observations. Since these factors (competition driving out competitors from the same area, hybridization swamping differences) do not occur in plants, isolation could not be a factor. There was little thought about the actual facts of distribution.

He explained why taxonomic and biogeographic considerations should play no role in considering evolutionary problems.

[T]he basal and underlying fault [for misunderstandings] consists in the fact that taxonomic and geographic methods are not in themselves, or conjointly, adequate for the analysis of genetic problems. [Genetic in the sense of origins.] The inventor did not reach the solution of the problem of construction of a typesetting machine by studying the structure of printed pages, but by actual experimentation with mechanisms, using printed pages only as a record of his success. Likewise no amount of consideration of fossils, herbarium specimens, dried skins, skulls of fish in alcohol may give actual proof as to the mechanism and action of heredity in transmitting qualities and characters from generation to generation . . . [60]

MacDougal went on to criticize a claim of Jordan's that *Oenothera Lamarckiana* might be a hybrid. He then turned to discuss Ortmann's attacks on the mutation theory. He took Ortmann to be antiexperimental, and chastised him for this stance.

Mr. Ortmann's discussions introduce a novel feature, in his estimate of the futility of experimental methods, which has the sole merit of boldness, coming at a time when the greater number of workers in the subject are turning from discussions and statements of opinion to actual observations. A mistrust is shown by him of experiments 'under artificial and unnatural conditions', as for instance in the botanical garden, or with domesticated forms.[61]

In response, MacDougal gave a lengthy discussion of the methods of "pedigree-culture." In this manner, through careful observation, genuine mutations could be discovered and studied. In these cases, "the variability . . . is not one of the modification of a character, but by total accession or loss of a character, and the variability is therefore a discontinuous one."[62] When the seeds from the mutant are separated, one can determine whether or not it was indeed a genuine mutation.

"The question therefore as to whether a plant is a continuous or discontinuous variant is one of simple measurement and estimation of qualities, not a matter of opinion."[63] For MacDougal, this was the way to discover the origin of forms.

In summary, MacDougal claimed that the methods of naturalists like Ortmann, Jordan, and Merriam were inappropriate to solve problems with regard to evolution, since evolution involves processes that can be discovered only through experimental means.

The problems included in a study of organic evolution are essentially physiological, and the elucidation of the mechanism and action of heredity by which qualities, characters and capacities are transmitted from generation to generation may be accomplished only by accurate observations and experimental tests with active or living material. The examination of preserved material not in hereditary series, or the wide generalizations derived from geographic studies, may not contribute to the progress of exact knowledge of genetics, or methods and manner of inheritance.

The combined and organized efforts of all the botanists in the world concentrated upon all the herbaria in existence would add but little to existing conclusions upon this subject, if we may judge by past achievements or immediate promise, while the most precise information upon geographical distribution can be of interest only in deciphering what has been accomplished, what forms exist and where, the factors influencing their movements, and where these have probably originated.[64]

Jordan responded to MacDougal in the September issue of *Science*.[65] He attempted to clear up possible confusions about the significance of the terms under discussion.

[T]he study of species and their relations to environment form the basis of the science of distribution. The species, as thus considered, is a kind of animal or plant as it has developed and as it appears in a state of nature.[66]

Jordan wanted the breadth of the geographical study of species to be made clear, as well as the recognition that species as they appear in nature were the things which need to be explained in an evolution-ary account.

Jordan claimed that facts about species in nature did not exhaust all the facts there were about them. "All species are plastic" and can change form "under new chemical or physical conditions." There-fore, "the field naturalist can not . . . know everything about any species, no matter how many individuals he may examine."[67] In this way, Jordan was able to show ways in which experimental study of organisms could be an important source of information.

At the same time, neither could an experimentalist working with breeding experiments or pedigree culture know all the facts about a species. The forms of these "garden naturalists" should not be as-sumed to be comparable to forms occurring in nature. On the whole, Jordan said, "[i]t is presumable that those naturalists know most about species as they are, who have given most time and thought to their study."[68] While other approaches are possible, the field natural-ist has a certain insight into plausible mechanisms involved in spe-cies formation.

> [I]t is fair to say that as the taxonomist of species finds in practically every case a geographical element in the development; as he finds that segregation and selection have apparently been accompaniments of nearly all changes in species, and as by these same agencies all species can be appreciably changed by the will of man, he may not unreasonably suppose that segrega-tion and selection have each taken some part in that life adaptation which we call organic evolution.[69]

Jordan went on to question the mutation theory as a general account of "species-forming". De Vries's notion that species go through "cy-cles of variation" is an "ingenious suggestion" but not really part of science.[70]

Thus, while Jordan accepted that "there is much – very much – about animals and plants, which can be learned only from experi-mentation under changed conditions," on the whole the methods of the naturalist – taxonomy, the study of geographical distribution, and so forth – were the key to understanding evolutionary processes.

> So far as species are concerned, it is clear that a large part of the problem demands the study of structure of forms and their relation to environment. There is much truth in Darwin's words that 'One has hardly a right to examine the question of species who has not minutely described many.'[71]

Jordan was concerned that experiments can be misleading if they are the sole form of evidence. In spite of this, biology became increasingly experimental to the detriment of other methods. By 1919, Jordan cautioned,

Our experiments detach a fragment of nature to be viewed in intensive detail; we succeed at isolating two or three of her minor problems, asking her to solve these for us without interference from the rest. We are not sure in these minor sections of nature that we have included enough or that we have taken too much to make the answers we receive intelligible or capable of rising to the rank of truths. An experiment is often the easiest line of attack, but it also may be the most deceptive.[72]

Jordan thought that in some sciences, like physics and chemistry, experimental "isolation" was relatively unproblematic, but in biology, the same techniques can be "relatively futile."

There were numerous responses by other naturalists, the most determined of whom was Ortmann. Ortmann responded to each of the attacks against Jordan, Merriam, and natural history more generally.

Ortmann criticized de Vries's view that he even derived species in his various experiments, since he never made any search of the terrain where he obtained his initial specimens to see if there were intergrades. In at least some cases (*Viola tricolor* and *lutea*) intergrades did exist. He also responded to MacDougal's charge that he was antiexperimental. "I have never said anything that might be construed as if I 'mistrusted' experiments or believed them to be 'futile', on the contrary I fully agreed that experiments ought to be made, but warned against too great complexity and improper interpretation."[73] Ortmann was not opposed to experimentation per se. His concern was that a great variety of conditions differ between the "botanical garden" where breeding experiments take place, and natural habitats, and that some of the differences in conditions could account for unusual effects like mutations. Therefore, different interpretations of these kinds of breeding experiments were possible, as opposed to the clear indication of underlying processes MacDougal saw.

Because of these difficulties, Ortmann suggested a cautious attitude toward experiments, and stressed the usefulness of fieldwork.

In my opinion, experiments should be made in close touch with nature, changing, if possible, only one or a few of the conditions, so that we may be able to record the effects of each single changed factor in the environment. But I do not believe that this is an easy task. On the other hand, we should bear in mind that nature has made and is making these experiments for us: the process of variation is going on continuously, and the effects of former variation are seen in nature, and may be studied in the shape of the actually existing variations, varieties and species, and their relation to the environment (ecology) . . . the modern ecological researches are just what is wanted.[74]

Ortmann similarly attacked Gager for reliance solely upon de Vries's experiments – and for his interpretations of those experiments. Ortmann claimed the experiments really showed the importance of isolation and selection on the part of the breeder.[75]

[H]e [Gager] points to the facts represented by the experiments, which I accept, but consider unsatisfactory and incomplete; and he points to the value of the experimental method as the only one that is apt to decide questions of evolution, which I positively deny. Experiments are valuable, but they should be properly understood, and should be correctly explained. The interpretation of his experiments given by de Vries is faulty, although the experiments themselves are indisputable facts; and the fallacy is due to his ignorance of the fundamental laws of evolution, and to his incorrect conception of the term species.[76]

Ortmann did not deny the validity or even the importance of experiments – but they needed to be properly understood, and they were not the only methods which could "decide questions of evolution."

CONCLUSION

I suggest that we should *not* interpret these conflicting statements as disputes over whether or not evolutionary questions required, or would benefit from, experimental research, since nearly everyone agreed that they would. Instead, this was more centrally a case of competing epistemologies. The arguments of Jordan and other natural historians utilized a variety of techniques, ranging from the taxonomic and biogeographic to the experimental and quantitative. They thought that the key to good science (or at least to good biology) was to obtain evidence for theoretical ideas from a wide variety of sources. This view contrasts with a *more* experimental epistemology which stresses the value of experimental control as the means of picking out true causes, and which therefore privileges experimental methods over all other methods of obtaining information. Vaughan, MacDougal, Gager, and Davenport all held that most of the *kinds* of evidence that the naturalists used were subjective and unreliable. Their view of how science should be conducted had no room for types of evidence that were not easily repeatable, since replicability would be an indicator that control had been achieved. As a result, very few lines of evidence could meet their standards and be used in support of their theory.

I have identified these differences in views about how science ought to be conducted as *epistemological* differences. They represent different views about how knowledge should be produced. These epistemologies are commitments to certain traditional scientific values or virtues as the key to successful scientific practice. The naturalists held certain virtues as central to science, especially consilience, the convergence of different lines of evidence in support of a theoretical view. In contrast, the mutation theorists stressed the virtue of experimental control. They considered this necessary in order for results to be objective, rather than mere "opinion."

De Vries, Gager, and MacDougal can be seen as advocating the replacement of other methods with the appropriate experimental techniques. Where the naturalist epistemology emphasized a number of lines of evidence (and hence a variety of techniques to acquire them), the goal of the "experimentalists" was to *replace* a variety of lines of evidence with one privileged line. Since this epistemology is held by many of the mutation theorists, these epistemological differences are reflected in the debate I have discussed.

Jordan saw the development of ecology by the 1920s as a continuation of many of the techniques of natural history. But he was concerned about the increasing specialization and concentration on genetics, which he saw as potentially misleading. Similarly, he found the low regard people had for taxonomy disturbing. Taxonomy was essential for the kinds of biogeographic evidence that he, Merriam, Grinnell, Ortmann, and other naturalists utilized in their work. The experimentalists who ignored all other kinds of information in favor of experimental evidence ran the risk that their experiments might be misleading or ignore certain other factors. Thus, in the dispute over the role of geographical isolation, the biogeographic evidence recognized by Jordan and the other naturalists seemed to indicate a problem for the mutation theory. At the same time, commercial breeders who failed to find mutants also began to oppose the theory.[77] Mutation theorists like Gager, who insisted that only replicable experimental evidence "counted," had been led to accept a bad theory – one which was to be discredited within a decade.

Speculation and analysis of chromosomal changes in plants[78] seemed to indicate that at least some potential examples of mutations (e.g., *Oenothera gigas*) could be attributed to polyploidy (doubling or otherwise increasing the number of sets of chromosomes) – and this was known not to occur in animals.[79] Jordan and other naturalists also speculated on the hybrid nature of some of de Vries's creations, denying that they were genuine cases of species production. Between 1910 and 1915, Bradley Davis (and others) showed through breeding experiments that some of de Vries's mutations were the result of the crossing of two heterozygotes: they were hybrids. Further studies showed unusual chromosomal behavior to be responsible for some of the peculiarities of de Vries's creations. "By 1915 the mutation idea, especially as it was put forth in its original form, had passed out of the serious biological community."[80] *Oenothera Lamarckiana* was an unusual plant, and hence the experiments de Vries performed were not a good way to capture experimentally the processes of evolution. The evidence, especially biogeographic evidence, available to the naturalists made it clear that there were problems with de Vries's theory and led them to conclude that evolution typically proceeded by geographical separation, followed by differentiation.

The views of the natural historians I have discussed had an important influence on those who developed the evolutionary synthesis. Mayr has consistently referred to the work of others when discussing the founder effect. He was influenced by Rensch, who was steeped in the literature I have discussed. In addition, Mayr was familiar with the work of Karl Jordan, D. S. Jordan, and Wagner, as well as with the work of Grinnell. Wright (whose views influenced Dobzhansky) was strongly influenced by Jordan and Kellogg. Kellogg's *Darwinism To-*

day (1909) was the work that introduced evolutionary theory to Wright.[81] This work discusses the ideas of Wagner, Rommanes, Gulick, and Jordan on the role of isolation in species formation. Wright cited a number of natural historians for their views on isolation and nonadaptive speciation.[82] Among them were Jordan, Kellogg, and Gulick.

Without the work of the naturalists whom I have discussed, a central theoretical part of the biological synthesis – our modern conception of speciation – would not have been developed. But although the naturalists' theories were adopted by the biological community, it was the epistemology of de Vries and other experimentalists that became dominant. The complaints Jordan raised in the 1920s could not turn back the tide. Although Jordan and his fellow naturalists won the theoretical battle over the role of isolation, they lost the epistemological war over what counts as "good science."[83]

NOTES

1. I would like to thank Jane Maienschein, Peter Galison, Tim Lenoir, John Dupre, Elihu Gerson, Michael Bishop, David Depew, and Julie Beeler Magnus for conceptual help and for reading through various revisions. An earlier version of this paper was presented to the International Society for the History, Philosophy and Social Studies of Biology.
2. Barber (1980), p. 28.
3. The most influential historical account embodying the approach I discuss is Allen (1978). For criticisms of this view, see Maienschein, Benson, and Rainger (1981). Many aspects of Allen's view are useful for understanding the case study considered here. However, on some readings, Allen cites the unjustifiably speculative nature of the naturalists' approach as a principal reason for the rejection of natural history. The case study presented here shows that the naturalists' science was more successful than they have been given credit for.
4. It is useful to think of these virtues as heuristics with normative force. See Wimsatt (1984) for an example of applying heuristics to scientific reasoning. One of the values of such a move is to illuminate the biases which attend any heuristic. It is beyond the scope of this paper to treat this issue in detail, but each of the heuristics or virtues I will describe in this case study has a corresponding bias, and in fact, the naturalists attempted to overcome the inherent bias in their own approach. By epistemology, I mean the package of virtues which are held by a community of scientists to be constitutive of good science (and hence the production of knowledge). See Magnus (1997).

5. Whewell (1847), pp. 173–4.
6. Ruse (1975), Thagard (1977). Not all scholars agree that Darwin's views were influenced by Whewell and Herschel. There are alternative explanations that could account for the structure of Darwin's theories.
7. In actual practice, there is a constant going back and forth between the two, so that the processes cannot be easily distinguished.
8. Herschel did not give this virtue a name. Laudan (1971), p. 55, has mistakenly incorporated this concept into the concept of consilience.
9. Anyone with the proper equipment and technical expertise.
10. Maienschein (1989), p. 21.
11. Maienschein (1989), p. 24.
12. Sulloway (1979). While Darwin's views on isolation are complex, he seems to have become convinced that isolated areas would contain too few individuals to support the amount of variation necessary for evolution. See p. 105 of the 1859 edition of the *Origin*.
13. Lesch (1975). The term speciation was not used until O. F. Cook coined it during this debate. However, the concept seems to have preceded the term.
14. De Vries (1905), p. 219.
15. Galton (1869), Provine (1971).
16. Bateson (1894), Provine (1971).
17. De Vries (1909–1910), v. 1, p. viii.
18. De Vries (1905), pp. 1–6; see also the significant front page, with quotations from Lamarck, Darwin, and de Vries, suggesting a progression in method.
19. Davenport (1905).
20. Jordan (1905), p. 547.
21. He sent the letter on Feb. 21, 1905.
22. That is, would one count such forms as species rather than subspecies, in which case question four becomes relevant.
23. This letter is provided on p. 548. A copy of the original is available in Jordan's letter books, Feb. 21, 1905, Stanford University Archives.
24. Stanford Archives, letter dated Feb. 21, 1905, marginal note.
25. Jordan (1905), pp. 552 and 550, respectively.
26. Jordan (1905), p. 559.
27. Jordan was not generally a neo-Lamarckian; however, he was unsure of the mechanisms responsible for variations and mutations, so he left open the possibility that the environment was responsible.
28. Jordan (1905), p. 550.
29. Merriam (1906), p. 242.
30. Merriam (1906), p. 243.
31. Merriam (1906), p. 247.
32. Merriam (1906), p. 251.
33. Merriam (1906), p. 254.

34. Vaughan (1906), p. 689.
35. Vaughan (1906), p. 687.
36. Vaughan (1906), p. 690.
37. Ibid.
38. Vaughan (1906), p. 683.
39. Gager (1906), p. 83, quote from De Vries (1905), pp. 790–1.
40. Gager (1906), p. 83.
41. Gager (1906), p. 84, quote from De Vries (1905), p. 465.
43. Gager (1906), p. 86, quote from De Vries (1905), p. 249.
44. Gager (1906), p. 86.
45. Gager (1906), p. 89.
46. Ibid.
47. Ibid
48. Davenport (1906).
49. Ibid., p. 557.
50. Ibid.
51. Ibid., p. 557f.
52. Ibid., p. 558.
53. Ibid., p. 557.
54. Kingsland (1991).
55. MacDougal (1906).
56. Ibid., pp. 207–208.
57. Ibid., p. 208.
58. Ibid., p. 209.
59. Ibid.
60. Ibid.
61. Ibid., p. 210.
62. Ibid., p. 213.
63. Ibid., p. 219.
64. Ibid., p. 220.
65. Ibid., pp. 223–224.
66. Jordan (1906).
67. Ibid., p. 399.
68. Ibid.
69. Ibid.
70. Ibid.
71. Ibid., p. 400.
72. Ibid.
73. Jordan (1919), p. 4.
74. Ortmann (1907), p. 187, and Ortmann (1906), p. 952.
75. Ortmann (1906), p. 952.
76. Ibid., p. 215.
77. Ibid., p. 217.
78. Kimmelman (1983).

79. Spillman (1907).
80. Actually, there are a few cases of polyploidy in asexual animals.
81. Allen (1969), p. 69. See also Cleleand (1972).
83. Provine (1986), p. 228.
84. Ibid., p. 226.
85. Although in many ways the so called "evolutionary synthesis" represented a return of consilience, there were many ways in which the practice of natural history was excluded from biology, and experimental and reductive approaches remained dominant. See, e.g., Vernon (1993).

REFERENCES

Allen, Garland. 1978. *Life Science in the Twentieth Century,* revised ed. Cambridge: Cambridge University Press.

Allen, Garland. 1969. "The Reception of the Mutation Theory." *Journal of the History of Biology* 2: 55–87.

Barber, Lynn. 1980. *The Heyday of Natural History.* Garden City: Doubleday.

Bateson, William. 1894. *Materials for the Study of Variation, Treated with Especial Regard to Discontinuity in the Origin of Species.* London and New York: Macmillan.

Cleleand, Ralph. 1972. *Oenothera: Cytogenetics and Evolution.* New York: Academic Press.

Davenport, C. B. 1905. "Species and Varieties." Review of De Vries, Species and Varieties. *Science* 22: 369.

Davenport, C. B. 1906. "The Mutation Theory in Animal Evolution." *Science* 24: 556–558.

De Vries, Hugo. 1909–1910. *The Mutation Theory,* 2 vols. Chicago: Open Court Publishing.

De Vries. 1905. *Species and Varieties.* Daniel Trembly MacDougal, ed. Chicago: Open Court Publishing.

Gager, C. S. 1906. "De Vries and His Critics." *Science* 24: 81–89.

Galton, Francis. 1869. *Hereditary Genius.* London: Macmillan.

Jordan, D. S. 1905. "The Origin of Species through Isolation." *Science* 22: 545–62.

Jordan, D. S. 1906. "Discontinuous Variation and Pedigree Culture." *Science* 24: 399–400.

Jordan, D. S. 1919. "Plea for Old-fashioned Natural History." Talk given at Scripps.

Kimmelman, Barbara A. 1983. "The American Breeders Association: Genetics and Eugenics in an Agricultural Context." *Social Studies of Science* 13: 163–204.

Kingsland, Sharon. 1991. "The Battling Botanist: Daniel Trembly MacDougal, Mutation Theory, and the Rise of Experimental Evolutionary Biology in America, 1900–1912." 82: 479–509.

Laudan, Laurence. 1971. "William Whewell on the Consilience of Inductions." *Monist* 55: 368–391.

Lesch, John E. 1975. "The Role of Isolation in Evolution: George J. Romanes and John T. Gulick." *Isis* 66: 483–503.

MacDougal, D. T. 1906. "Discontinuous Variation in Pedigree Cultures." *Popular Science Monthly* 67: 207–224.

Magnus, David. 1997. "Heuristics and Biases in Early Evolutionary Biology." *Biology and Philosophy* 12: 21–38.

Maienschein, Jane. 1991. "Epistemic Styles in American and German Embryology." *Science in Context* 4: 407–427.

Merriam, C. H. 1906. "Is Mutation a Factor in the Evolution of the Higher Vertebrates?" *Science* 23: 241–257.

Olby, Robert. 1985. *The Origin of Mendelism.* Chicago: University of Chicago Press.

Ortmann, A. E. 1906. "Facts and Theories in Evolution." *Science* 23: 947–952.

Ortmann, A. E. 1906. "The Mutation Theory Again." *Science* 24: 214–217.

Ortmann, A. E. 1907. "Facts and Interpretations in the Mutation Theory," *Science* 25: 185–190.

Provine, William B. 1971. *The Origins of Theoretical Population Genetics.* Chicago: University of Chicago Press.

Provine, William B. 1986. "Sewall Wright and Evolutionary Biology." In David L. Hull, ed., *Science and Its Conceptual Foundations.* Chicago: University of Chicago Press.

Ruse, Michael. 1975. "Darwin's Debt to Philosophy." *Studies in the History and Philosophy of Science* 6: 159–181.

Chapter 6

Competing Epistemologies and Developmental Biology

JANE MAIENSCHEIN

In 1926, American embryologist Herbert Spencer Jennings reflected on developments over the previous decade. He recalled the 1890s, when, in a spirit of great enthusiasm for experimentation, one after another embryologist did experiments, got results, and drew different conclusions. Indeed, they often drew *quite* different conclusions. It reminded him, he reported, of a Gilbertian comic opera [particularly the *Mikado*] in which all the characters claim success. They happily sing, "For I am right, and you are right, and he is right and all are right." Despite their apparent disagreements and contradictions, all are perceived as "right." (Jennings 1926: 99) The issues are not always about the concrete details of evidence or the niceties of theoretical interpretation. They are not even always about having one "winner" in a given case. Instead, the issues often hinge on rightness in a different sense. Often the issue concerns what is to count as evidence or how much certain evidence is to count for or against a given argument. In short, the question often concerns what counts as the "best science" within the constraints and context of the community at the time.

At root, these are issues of epistemology. Claims of rightness or bestness carry with them views about knowing: what does it mean to be right? What counts as evidence in favor of claims to being right? How do we know? This is not to say that the cases Jennings was recalling did not also involve disagreements about precisely what the researchers saw, or about how best to acquire data or evidence. Nor did the cases exclude arguments about theoretical interpretations or metaphysical convictions. Indeed, historians as well as the biologists themselves have often emphasized the empirical evidence

122

or the theories or details of the experimental design. Philosophers and historians have considered whether theory or the scientific practice that produces data takes precedence, and how scientists have negotiated the relations between theory and practice, while sociologists have considered the ways that communities allow selected theories and practices to become conventional and thus established for that group.

What biologists have generally missed, and what historians and philosophers have often tended to misunderstand, however – even during the recent resurgence of enthusiasm for thinking about such things as "social epistemology"–is that epistemology actually can drive the science. Epistemological issues of rightness, and the coexistence of competing sets of epistemological values, often strongly direct the scientific discussions and underlie the controversies involved in particular cases. It is the epistemic norms, after all, that say how scientists should select their data, evaluate their experiments, and judge their theories. It is epistemic convictions that dictate what will count as acceptable practice and how theory and practice should work together to yield legitimate scientific knowledge. It is epistemology that underlies consideration for a given group in a given time and place of who is right. And competing epistemologies can coexist in a scientific community – and fruitfully coexist.

Let us begin with three parallel examples from what we now call developmental biology and genetics, and move on to other types of cases. I will present each case in stark outline, extracting key features of the discussion and ignoring much of the potentially rich contextual discussion. That detail appears elsewhere, and what I want to do here is to draw on texts to show how epistemology matters. After discussing the cases themselves, I will consider what they tell us about how science, and scientists, work.

PREFORMATION AND EPIGENESIS I: WOLFF AND BONNET

Enter Caspar Friedrich Wolff (1759) and Charles Bonnet (1769). Participants in the lively discussions about embryonic development during the mid eighteenth century, this Russian-turned-German and the Swiss scientist provide clear alternatives. They looked at the same thing, and at some basic level they agreed in their description of what they saw, yet they drew quite different conclusions. They provide an apparently clear case – but a case of what?

Sometimes their story is presented as a case of alternative theories. Wolff was an epigenesist, who held that development of each individual embryo proceeds from an unformed and basically homogeneous state to the fully formed adult stage. Development, for Wolff, was gradual and progressive. By contrast, Bonnet and his counterparts held a preformationist view. Form does not emerge gradually from nonform. Rather, the form must have been there all along in some preexistent, preformed state. On the face of it, epigenesists and preformationists were arguing about a particular theory.

Yet on another axis lies metaphysics, and they also disagreed about that. In general, though not necessarily, epigenesists were vitalists during the eighteenth century and for most of the nineteenth century as well. It was very difficult for an epigenesist to explain the emergence of form if it did not come from somewhere. Vitalism provided a source. Therefore, those who began with epigenesis tended to move also to vitalism: epigenesis first, then vitalism. By contrast, materialists were led to preformation. How else to explain the existence of form – highly differentiated and highly specific form?

Shirley Roe has beautifully developed this discussion of materialism and vitalism, epigenesis and preformation, for a range of eighteenth-century principal players, and I need not repeat that discussion here or discuss all the myriad other players in the debates (Roe 1981). Rather, let us note something important. While the two positions differed on metaphysics and theory, they agreed on the goal of achieving an explanation of the phenomena. They even agreed on the phenomena. But they had much different views about the epistemic value of their observations.

Both Wolff and Bonnet studied chick development, and both looked closely at the twenty-eight-hour stage – shortly before the heart becomes clearly visible and clearly beating. They had no way to observe the process of development, no special secret window into the egg to observe every moment of the progress. They had to take what were, in effect, snapshots frozen in time and extrapolate by making assumptions about what happened in between. Furthermore, they had to do this with different individuals, since once they had taken the picture that particular embryo no longer existed.[1]

Wolff looked at his twenty-eight-hour chicks and said, basically, I don't see the form or all the parts, I don't, for example, see a beating heart. Therefore it is not there. It could be that our senses just aren't good enough, but then we should see more parts with a more powerful microscope lens. That does not happen, however; the lens just makes the existing parts clearer and bigger. For Wolff, he does not see it, and therefore it does not exist. Seeing is believing – and not seeing is believing not. This is a powerful epistemological conviction about how we should go about believing – and knowing – things.

Bonnet had equally strong epistemological convictions. He looked at his twenty-eight-hour chick and could not see the form of all the parts either. He agreed that all the parts were not yet visible. Unlike Wolff, however, he insisted that they *must* be there. We know that form exists later, and we know that we need an explanation of how it got there. Furthermore, that explanation must be in terms of matter and motion, or materialism.

Note that the need for a valid explanation was as important as the need for materialistic roots or for a preformationist result. Indeed, if Bonnet had been willing to start with materialism and the later existence of form and to say, "I do not know how the form arose. Hypotheses non fingo," he would not have needed preformation. If Wolff had been willing to start with empirical evidence about that form and to say, "I do not know how the form arose. Hypotheses non fingo," he would not have needed to invoke a vitalistic something that he could not directly see. The insistence on having an explanation is the epistemological conviction that the possession of such an explanation is what constitutes knowing and produces good science. Bonnet would conclude that not seeing is not determinate. The issue is, in part, how much we know and what we know from what we see. The epistemological forces leave Wolff with his epigenesis and vitalism and Bonnet with his materialism and preformation. Each believed he was right. And there was room for the coexisting competing epistemologies, even though eventually the scientific community, while endorsing a general empirical disposition to think that seeing should be believing, decided that accumulating evidence favored Wolff's epigenetic interpretation and Bonnet's materialistic metaphysics.

EPIGENESIS AND PREFORMATION II: ROUX, DRIESCH, AND FRIENDS

The second example takes us to the end of the nineteenth century with a parallel debate. Two German embryologists, Wilhelm Roux (1888) and Hans Driesch (1891), play the lead roles here. In this case, rather than starting with the same observations and developing different theories, they began with the same theory, performed what they regarded as fundamentally the same experiment, made different observations, and drew quite different conclusions. They each quite confidently drew those divergent conclusions. And they diverged for epistemological reasons.

Roux laid out the theory: embryonic development is mosaic-like, that is, each cell division divides the originally inherited complement of "determinants" that cause cell differentiation into the separate cells (Roux 1888). This was essentially a preformationist, or rather a modified predeterminist, view. Roux predicted that if a researcher could remove one of the two cells after the first cell division, the result should be a half-embryo. Only half of the original determinants would persist to yield this half-embryo. One and a quarter centuries after Wolff and Bonnet, Roux recognized that passive observation alone would not get the embryologist very far and urged the use of experimentation to produce additional observations and to control and test ideas.

To test his prediction experimentally, Roux took the frogs' eggs that were readily abundant in his area and killed one of the two cells (blastomeres) after the first cell division. He used a hot needle and observed that this blastomere failed to develop further; it just stayed there as an undifferentiated lump. As predicted, Roux concluded that the remaining blastomere developed as it would have under normal conditions and yielded, in effect, a half-embryo. Therefore: a brilliant confirmation of the original hypothesis using a well-designed test. He triumphantly declared the mosaic theory to be correct.

The slightly younger Driesch was inspired by Roux and believed that Roux was right – in his interpretation and in his approach (Driesch 1891). He resolved to take the same mosaic theory, to make the same prediction, and to perform the same experiment. But while Roux had abundant frog eggs, Driesch worked at the Naples Zoological Station and had abundant sea urchin eggs. It also occurred to him that sea urchin eggs might even be better for getting clean results, since the Hertwig brothers, Oscar and Richard, had shown at Naples that it is possible fully to separate the blastomeres by shaking sea urchin embryos after the first or even second cell division. Thus, Driesch could obtain isolated cells where Roux had had to settle for killing half the material, which remained inertly attached to the still-living cell.

Driesch's theory, prediction, experiment, and basic approach were all the same as Roux's. And he recorded, "I must confess that the idea of a free-swimming hemisphere or a half gastrula open lengthwise seemed rather extraordinary. I thought the formations would probably die." (Driesch 1891: 46) But not so. "Instead, the next morning I found in their respective dishes typical, actively swimming blastulae of half size." Indeed, what Driesch got was two small forms. According to prediction, this was not at all right.

Driesch therefore concluded that the predeterminist mosaic theory must be wrong. Instead of cell differentiation because of the partitioning of inherited determinants at cell division, Driesch postulated that the cells must each retain some "totipotency" (or potential of the whole). They have the ability to undergo internal regulation in response to the needs of the whole. The cells, therefore, behave by working together as a "harmonious equipotential system." Driesch accepted his new observations with sea urchins as constituting new evidence and positive knowledge, which led him to reject the proposed theory and to develop a new alternative explanation. That, after all, is how the experimental approach, much acclaimed by Roux himself, is supposed to work.

It might seem that Roux must surely follow Driesch and admit Driesch's rightness in the face of such a powerful counterexample to his own theory. He did not. Nor did Roux develop an alternative theory of his own. Rather, he stuck with his mosaic theory, saying that it was still right. Surely that was bad science on his part.

But no, Roux invoked additional values. Science seeks explanations, he urged, in the form of explanatory theories. The mosaic theory is such a valuable theory, he persisted, explaining so many different things and providing the best theory available. All we need is a little auxiliary hypothesis to fix things. To this end, he postulated the existence of a "reserve germ plasm." Aha: so the original germ material and its determinants get divided up into cells in the normal division process. Yet in some cases for some organisms, there is a backup set of determinants to step in and carry on the process. Roux did not actually physically see such a reserve; he had no direct empirical evidence for its existence. Yet he believed that postulating its existence was justified, and indeed necessary, since the theory was so clearly important for producing knowledge in the form of materialistic explanation – which is what he valued most, and what made his science right. He sought to save the theory even in the face of apparently contradictory phenomena.

On the face of it, Driesch's is a clearer experiment, since it actually separates the blastomeres completely. Driesch's epistemology told him to accept those experimental results and to revise the theory – even though this later pushed him toward a vitalistic theory for reasons like those that had made Wolff a vitalist, namely, the need to provide an explanation of the emergence of form from the unformed. Roux's epistemology told him to go with the good theory apparently capable of explaining so much and so many different kinds of phenomena of heredity and development. Each believed he was right.

Yet these two were not alone. Others joined the discussion and sought still further evidence, with further experiments and more reflection on the interpretation of results. The American cytologist Edmund Beecher Wilson was one of these.

Wilson was intrigued by the difference between Roux's and Driesch's results and sought to understand how such apparently different results could occur and how best to interpret them. Wilson said, quite reasonably, that we need to seek answers with additional experimental evidence – with different organisms and different situations to control more factors. He used nemertine eggs (*Cerebratulus lacterus*) and others, since "[I]t is obvious . . . that this question is one not for speculation but for further experiment" (Wilson 1906: 265–266). The result was some of both patterns, which led Wilson to conclude that development is more complex than either Roux or Driesch had recognized. For Wilson, any satisfactory explanation, anything that could possibly be accepted as knowledge, must take that complexity and diversity into account. The researcher should move carefully from observation to conclusion. Wilson realized that seeing does not simply lead to believing, nor does a great theory carry the day simply because it can apparently explain more or has more immediate promise.

In retrospect, Wilson's approach and his cautious conclusions make a lot of sense. Yet note the underlying assumptions. Wilson rejects an emphasis on any one model organism and its clear and compelling theory, or on any one crucial experiment. His approach requires waiting for more evidence – and how much is enough? He has no theory to tell us that. He deemphasizes the role of theory and of explanation in favor of accumulating more data. Roux says to follow the theory and its explanatory and predictive power. Driesch calls for following the compelling experimental discrepancy to a new theory. Wilson calls for continuing to accumulate more evidence.

They were arguing not just about form and what causes organization of form, but also about how to study it. These are, at root, fundamentally different epistemological values addressing what matters most in science. And there is no way any of these man could have persuaded the others of his rightness, though Wilson and Driesch certainly tried in an extended correspondence. Interestingly, in retrospect Roux is often praised by biologists for his invoking of the "modern experimental method" for biology, even though Driesch's epistemological approach and his following the empirical evidence conforms better to the description of the modern approach. And Wilson, the careful researcher, has been forgotten by all but a few historians and older biologists (Maienschein 1991).

EPIGENESIS AND PREFORMATION III: NERVES

A third epigenesis–preformation case, also from the late nineteenth and early twentieth centuries, shows the variety of ways that superficially similar debates can involve quite different underlying issues. The question centered on how nerves develop. Since nerve fibers play an obviously important role in making complex functional neural connections, researchers began to ask how they do that. How do fibers "know" how to make the proper connections? Or do they "learn"? In other words, is the connection predelineated or pre-established in some material way, or does it emerge and find its path only gradually, epigenetically over time and in response to whatever formative forces are operating? Further, do the individual neurons and nerve fibers act and grow independently, or do they make up an integrated nerve net, in which the cells may even interconnect into a reticulum? The Italian researcher Camillo Golgi and the Spanish investigator Santiago Ramon y Cajal played central roles in this debate. Golgi argued for reticular nets, Ramon y Cajal for autono-mous neurons.

Golgi and Ramon y Cajal both looked at killed and prepared neural material. They used essentially the same methods and, in-deed, some of the same specialized techniques including "Golgi preparations" with silver nitrate impregnations and advanced stain-ing. At least early in their work, they apparently respected each other's technical abilities and even referred to each other's prepara-tions as "evidence." (Their later battles may have had more to do with establishing themselves as deserving primary credit and ulti-mately the Nobel Prize – which they eventually shared – than with their deeper convictions about how best to do science.)[2]

What differed significantly from the beginning was which exam-ples they regarded as important for understanding the development of form. They had different views about which phenomena were really "data" and "evidence," about which observations should "count," as well as about how and when the nervous system is organized. Yet they each kept gathering more of the same kinds of observations and largely ignoring or discounting the other's. Golgi selected examples and worked to create more examples that clearly show the apparent interconnections and the nets, while Ramon y Cajal selected and developed examples to reveal apparently separate cells. They could have spent more effort commenting on why their own selections made better material, or on what was wrong with the other's selections, but at least at first they largely made this an im-plicit matter. Each made his selections and argued, dancing around with increasing vehemence and eventually with significant vitriol, that his own selections were right.

The situation seemed irresolvable, with fundamentally incompatible assumptions, until other participants entered with alternative approaches, different focuses, and still further competing epistemologies. American embryologist Ross Harrison saw the difficulties both in selecting which material to count and in interpreting the dead and manipulated preparations that were far removed from their original living state. He sought a way to achieve what could be generally accepted as "definitive knowledge." Inspired by the experimentalism, with its promise of control and respectability, of Roux, Gustav Born, and others, Harrison worked to devise an experimental test of the theories about how nerves develop. He maintained that performing an experiment with actual living, developing neural material would be better than relying only on the preserved specimens of Golgi and Ramon y Cajal – better for producing reliable knowledge in the form of both observations and explanations (Harrison 1910).

He first assumed that since nerve fibers are the outstretching parts of nerves that connect with other nerves, it would be legitimate to focus on nerve fiber growth. Next he assumed that a key question was whether nerve fibers can develop independently and separately. If they are capable of doing so, he argued, then it is reasonable to assume that they can do so in normal conditions – and that they normally do. If they can do so, there can be no legitimate argument that it is necessary to have preexisting nerve nets to guide or make possible normal nerve development. Thus, using small nerve cells transplanted into drops of frog lymph in the first successful tissue culture, he devised what he regarded as a "crucial experiment" to determine whether fibers can grow independently by protoplasmic outgrowth. They did, therefore they can.

Harrison's approach assumes that his artificial, experimental, highly contrived conditions will yield information useful for understanding normal development. It assumes that what happens in this artificial controlled setting parallels what happens in the living organism. It assumes that such an experimental approach yields reliable and warranted knowledge.

Perhaps astonishingly, others agreed. Even many earlier critics of the neuron theory with its independent nerve cells came to agree that Harrison had provided "proof" of the theory, with its epigenetic implications. This did not happen overnight, and it required a campaign over several years and with increasingly sophisticated experimental design, but it worked. Though we know in retrospect that proof is a complicated thing, at least for his contemporaries Harrison was regarded as right. And he seems to have carried the day by pursuing a set of epistemological values and an experimental approach that won over the scientific community that came to endorse similar views.

VARIATIONS ON A THEME: MORGAN

Turning to yet a different case, we find the American Thomas Hunt Morgan apparently at odds with himself. This is a case where an essentially consistent epistemology underpinned quite distinct theories and research approaches, and a case where one researcher shifted from an epigenetic to a more preformationist position, for epistemological reasons.

Prior to 1910, Morgan had been studying development, especially focusing on regeneration in a wide range of organisms (which he saw as a kind of "natural experiment"). As he wrote to a friend in 1908, "my field is experimental embryology." He viewed heredity as much less interesting, basically serving to insure stability by making offspring much like parents, as more conservative and more preformationistic than he could accept. Development, in contrast, was for Morgan an interesting and creative process that produces variation and brings the process of developmental mutations in a more epigenetic way that he felt best fit the facts.

Morgan was looking for de Vriesian mutations in various organisms, including the fruit fly *Drosophila*, as a source of variation. He wrote to a friend that he was about to give up on the messy annoying flies (and the requirement to provide rotting bananas for them to feed on), but he did not. He found a peculiar white-eyed male, where all others had red eyes. The famous resulting Mendelian ratios of offspring's sex-related traits suggested strongly that heredity both operates in a Mendelian way and is connected with the chromosomes. Morgan did change his emphasis and enthusiastically took up heredity and *Drosophila* as his primary research program (Maienschein 1991, chapter 8; Morgan 1910).

That is a familiar story, and it is often recounted to show how Morgan changed his mind on the face of empirical evidence. It is suggested that, in effect, the previously misguided Morgan herewith saw the error of his ways and corrected his approach. This is offered as a story about the triumph of empiricism and reduction, and of experimentalism, over his former less defensible developmental views. It is also often offered as a tale of the triumph of the Mendelian-chromosome theory, with its message of genetic determinism and its first step toward our present enthusiastic search for "the gene for [whatever trait]." It was partly that – but only partly.

Yes, Morgan did change his mind and embraced a Mendelian-chromosome theory as a strong interpretive framework from which to explore further the actions of heredity and development. Yet Morgan insisted that science does not work in such a black-and-white way. It is not that he had seen the error of his former ways and, in a grand revelation, finally embraced the truth. Instead, he explained, "I beg to remind the reader, and possible critic that the writer holds all conclusions in science relative, and subject to change for change in science does not mean so much that what has gone before was wrong as the discovery of a better strategic position than the one last held." (Morgan 1913: iv).

The particular theory was not very important to Morgan, and served effectively as the temporarily best working hypothesis instead of as some capturing of reality. The experimental evidence was important, of course, but of only passing interest. What was most important about the evidence was the way it weighed in favor (or against) a theory. Morgan's deepest commitments were neither evidentiary nor theoretical but epistemic. His only enduring commitment was to his epistemological standards for what counts as good science.

Morgan did *not* change his mind about how to do science. All along he said that researchers should pursue the best theoretical interpretation most consistent with available data and most capable of producing further knowledge. He embraced an experimental approach, and he was fundamentally – and productively – an opportunist who pursued the "better strategic position" for any given time and context. This led him, quite naturally and logically, to change his mind about the Mendelian-chromosome theory of heredity.

Morgan's rightness lay in his solid and persistent epistemological convictions – even while that led him to different research questions, approaches, and interpretations over time.

COMPETING EPISTEMOLOGIES

There are many more examples that show the way individual disagreements play out, but let us look at a few types of cases to illustrate the range. One centers on what has been termed the naturalist-experimentalist or field versus laboratory debate. Though these are different debates, what is at issue in both is how much we can learn in the laboratory – by extracting life from its natural ecological setting and seeking to contrive and control conditions, by preparing, slicing, dicing, poking, measuring, and such. Or can we learn more – and perhaps learn better–by studying the messy, muddled life-in-its-context? Assuming that biology is the study of life, which approach is better? Which allows researchers better to know about life? Which epistemological conviction about how to do good science wins? Which is right?

One of my colleagues, Douglas Chandler, uses electron microscopy to study cells. He has developed a very useful freeze fracture technique that freezes the cells, then fractures them in a way such that they break along the "fault lines" corresponding to internal structures within the cell. This method has proven invaluable for gaining insight into the internal workings of the cell. With time, he and others have developed revised approaches including ways of freezing the cell more quickly so as to produce less damage and distortion, such as with deep etching techniques. Yet the entire approach requires killing the cell and observing it under artificial and experimental conditions. Those who insist on studying life in its living, functioning, active form reject the entire approach and its epistemological assumption that we can indeed learn about nature from such controlled and contrived conditions. Meanwhile, Chandler insists that there are things that we can learn from such techniques, things which constitute important contributions to knowledge, that we cannot learn otherwise. Which approach is right? And which should receive the funding, given limited resources and intense competition for the rewards?

In the laboratory, then, what techniques and approaches provide the best science? With cytology: is it better science to study cells physiologically, actively, to study processes of development? Or is it epistemically acceptable to take a snapshot of killed and preserved materials, seeking to gain knowledge about the morphological structure of the cell? Often the different approaches produce different results and conflicting interpretations. Which is right?

Is it the systems ecologist who looks at the dynamically interacting components of a system who is right, or the evolutionary ecologist who insists on the historical, evolutionary features of each adaptive unit? Is it the geneticist who seeks genetic determinants as providing knowledge about the cause of traits, or the developmental biologist who insists on historical process in any explanation of biologist traits? Each of these positions brings a competing view about appropriate epistemological values, and there are innumerable parallel examples that illustrate the range of ideas about "rightness."

Perhaps this is a case of Gilbertian rightness, and everyone can dance about singing happily of his own rightness. After all, diversity is considered a virtue these days. Why not accept all the competitors; what possible criteria can there be for adjudicating among the various competing claims? Perhaps none definitively, but we should try. Some science will be selected – for example by NSF, NIH, or those doing hiring – as better than other science. Some will be funded and some will be published, but much will not. The stakes are high. It is worth thinking about why, about who is more right at any given moment and according to what standards.

The cases show that epistemology matters and that competing sets of epistemological norms can coexist and make science lively, exciting, and perhaps even more productive as more research appears attempting to resolve the debates. We have seen a range of cases involving views about epigenesis and preformation, about heredity and development, about materialism and vitalism, about different theories and different practices, and about how and where best to study life. We have different views about what we are looking for in science, about what will count as data and evidence, as legitimate patterns of inference – in other words, as ways of knowing and of gaining knowledge.

We have cases of seeing the same thing differently, of weighing the

evidence differently, of seeking different bits of evidence and counting theory more heavily than empirical evidence, of counting experiment or field observations as most important, of valuing explaining
or more data or experimental testing as most important, of preferring
"natural" to "artificial" or controlled conditions.

These are cases where epistemological views are central, cases
that together illustrate the richness of science, the value and legit-
imacy of coexisting competing views, and the way such debate can
be good for science. They tell us a lot about how science works and
about the range of what should count as "right" or "good."

Furthermore, developing such examples and their implications
gives us something desirable: a better picture of how science works
and a way of discussing what should count as "good science" that is
research-based and data-driven, that is intellectually and historically
defensible, and that is useful. We can leave it to others to develop this
usefulness, or we can do it ourselves by trying to sort through ways
to authenticate, even if not to adjudicate, the competing claims to
rightness.

Yet this does not mean that "anything goes," nor that there is no
basis for judgment. There are constraints on the values of any community, and these come from conventions of the community. The
rational epistemologist would allow discussion and debate. Such an
epistemologist would smile at the dancing in the square as all are
convinced that they are right, and would then allow us to work at
adjudicating the competing claims based on reason and the existing
values of science. Those in the scientific community who embrace
the reason and logic of the enlightenment should agree. And once
we agree, we can comfortably endorse the coexistence of competing valid epistemologies within the scientific community at any
given time and place, and for any group of researchers. Then we can
get down to the work of understanding how to make justifiable
demarcations and just how much and in what ways epistemology
matters.

ACKNOWLEDGMENT

Thanks to the National Science Foundation for support of research for this
project.

NOTES

1. For an accessible look at Wolff's and Bonnet's central ideas, see Hall (1951), pp. 371–372, 377–381.
2. There are many discussions of Golgi and Ramon y Cajal's works. For useful overviews see Clarke and O'Malley (1968), pp. 91–96, 109–113; Brazier (1988), pp. 143–144, 145–146; Maienschein (1991), pp. 268–293.

REFERENCES

Brazier, Mary. 1988. *A History of Neurophysiology in the 19th Century.* New York: Raven Press.

Clarke, Edwin and C. D. O'Malley. 1968. *The Human Brain and Spinal Cord.* Berkeley: University of California Press.

Driesch, Hans. 1991. "The Potency of the First Two Cleavage Cells in Echinoderm Development." Translation in Willier and Oppenheimer, 38–59.

Hall, Thomas S. 1951. *A Source Book in Animal Biology.* Cambridge: Harvard University Press.

Harrison, Ross. 1910. "The Outgrowth of the Nerve Fiber as a Mode of Protoplasmic Movement." *Journal of Experimental Zoology* 9: 787–846.

Jennings, Herbert Spencer. 1926. "Biology and Experimentation." *Science* 64: 97–105.

Maienschein, Jane. 1991. *Transforming Traditions in American Biology, 1880–1915.* Baltimore: Johns Hopkins University Press.

Roe, Shirley. 1981. *Matter, Life and Generation: Eighteenth-century Embryology and the Haller-Wolff Debate.* Cambridge: Cambridge University Press.

Roux, Wilhelm. 1888. "Contributions to the Developmental Mechanics of the Embryo." Translation in Willier and Oppenheimer, 2–37.

Willier, Benjamin H. and Jane M. Oppenheimer. 1964. *Foundations of Experimental Embryology.* Englewood Cliffs, N.J.: Prentice-Hall.

Wilson, Edmund Beecher. 1904. "Experimental Studies on Germinal Localization." *Journal of Experimental Zoology* 1: 1–72, 197–268.

Chapter 7

From Imaging to Believing

Epistemic Issues in Generating Biological Data

WILLIAM BECHTEL

> Visual displays are curiously robust under changes of theory.
> You produce a display, and have a theory about why a tiny
> specimen looks like that. Later you reverse the theory of your
> microscope, and you still believe the representation. Can the-
> ory really be the source of our confidence that what we are
> seeing is the way things are?
>
> Hacking 1983: 199

"Seeing is believing." Or so we are often told. But we don't have to go far to find examples in which seeing offers insufficient grounds for believing. Most people, looking at Figure 7.1, make the perceptual judgment that the two shaded parallelograms are not identical. However, if you were to cut out one parallelogram and place it on top of the other, you would convince yourself that they are indeed the same. But even acquiring the belief that they are the same will not change how you see the two parallelograms or the perceptual judgment you make, for they will continue to look to be of different shapes. If we think of the visual system as an instrument for producing perceptual judgments, then we would construe the judgment resulting from viewing this figure to be an artifact. Fortunately, our visual system is quite well adapted to the environment in which we live; accordingly, such illusions are rare, and our perceptual judgements do provide good ground for belief.

Increasingly, scientists rely on specially designed instruments to produce the perceptual inputs that result in perceptual judgements. Like the visual system itself, these instruments often produce images that are very compelling. Nonetheless, when these instruments are

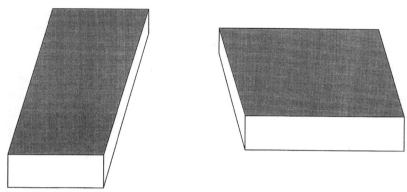

Figure 7.1. Illusory figure after Gaetano Kanizsa. The two shaded parallelo-grams are identical in shape and size, but rotated.

first put to use to generate images in science, the community of scientists often questions whether these images and the perceptual judgments they yield ought to compel belief. Many will claim that in fact the instruments are producing artifacts – that the structure produced in the image does not reflect the structure in the phenomenon that it is claimed to reflect but is due to the instrument. Indeed, the controversies surrounding such instruments can be among the most bitter in science. Eventually, they dissipate: procedures are developed for using the instrument in ways that the community agrees are producing reliable information about the structure of the phenomena. These procedures are widely disseminated, and the community comes to accept the data produced from the instruments. They become, in the language of Latour (1987), black boxes.

In this chapter I will focus on the introduction of two different techniques for producing visual images in modern biology: electron microscopy and positron emission tomography (PET). These techniques were introduced into biology during different periods (electron microscopy in the 1940s and 1950s, and PET in the 1980s and 1990s) and are used to render imagable different things (structures in systems versus functions of components of systems), but each initially generated controversy as to whether it was in fact generating artifacts. My objective is to identify the sources of epistemic controversy raised by the use of the new instruments and the ways in which scientists have gone about settling these controversies. I will begin by describing briefly the introduction of the new instruments and drawing out what it was about the way each instrument was used that raised the possibility of artifacts. I will then shift to our own visual system and show that the sources of controversy are present there as well. We can then consider how we resolve controversies about perception when only our visual system itself is involved. This will in turn provide the framework for examining how the controversies over these two biological instruments have been resolved by scientists.

INTRODUCING NEW INSTRUMENTS IN BIOLOGY

Both electron microscopy and PET are based on principles developed in other areas of science. For the most part, these principles were not what was in contention when the techniques were introduced into biology. Rather, what generated controversy were the procedures required to employ the instruments to secure data about biological phenomena. With both electron microscopy and PET, there were numerous reasons to be suspicious that these procedures were creating artifacts. First, many steps are involved in preparing biological material to be imaged, each of which could alter the phenomenon so that the final image reflected the alteration, not the original phenomenon. Second, there is a great deal of flexibility in the ways these procedures are applied, which is often revealed in differences in the images generated by different laboratories. Finally, researchers often do not understand in detail how their interventions alter the phenomena and so do not know whether they are introducing distortions.

The Electron Microscope[1]

The electron microscope and the theory underlying it were developed around 1930, and the microscope was put to use in imaging physical structures (Marton, 1968). Accordingly, the basic questions about the operation of the instrument (e.g., the techniques for generating and focusing the electron beam) and the generation of images using it had been largely resolved before it was applied to biology in the mid 1940s. There was powerful motivation to apply it to biological specimens. The limit of light microscopy stemmed from the wavelength of light, which set absolute limits on the resolution that could be obtained (2,500 Å or .25 µm). While the basic structure of the cell and major organelles such as the mitochondrion were large enough to be viewed with the light microscope, given appropriate fixatives and stains, any finer structure in the cell, especially those internal to such organelles, could only be visualized with a new tool. The electron microscope promised to provide that tool. The wave length of an electron is dependent upon voltage, but even the relatively low-powered 50 kV microscopes developed in the 1930s and 1940s theoretically permitted resolution down to 5–10 Å (50–100 µm). (For a discussion of how cell biology developed as a scientific discipline, partly as a result of the advent of the application of the electron microscope, see Bechtel 1993.)

But there were serious obstacles to be overcome before the electron microscope could be put to use.[2] First, the electron beams of 60 kV microscopes would only penetrate specimens of 0.1µm in thickness, which is much thinner than the typical cell. Second, any specimen for electron microscopy had to be placed in a vacuum. Finally, since most biological material contains approximately the same amount of matter and hence does not differ in ability to scatter electrons, some means of contrast enhancement was required.

Solving each of these problems required altering the biological specimen, which raised the possibility of an artifact. As we shall see, there were often alternative ways to solve each of these problems. When very different procedures result in consilient results, researchers acquire an epistemic check on their procedures. But demonstrating such consistency is not always easy. For example, researchers developed two very different ways of achieving thinness: tissue culturing cells and slicing cells with a microtome. Tissue culture involves growing cells on a glass surface in an artificial chemical medium and outside the three-dimensional matrix of the organism. The result is that cells spread out more thinly than usual; at the boundaries where they spread most thinly they can be penetrated by the electron beam. The alternative approach of slicing cells thinly had already been developed for light microscopy, but at that time the microtomes could only prepare specimens that were approximately

1μm in thickness. The challenge in generating yet thinner sections was to move the knife through the cell without tearing any of its constituents. A number of new microtomes were developed around 1950, as well as techniques for gently heating the cell so as to advance it toward the microtome blade (Cosslett 1955). While they do so in different ways, both tissue culture and microtomy involve treatments that could distort the internal structure that electron microscopy is intended to reveal. Figure 7.2, to which we shall return, illustrates how different the results of the two techniques could appear and the challenge involved in establishing that they were showing the same thing.

If producing a sufficiently thin specimen for the electron beam provided serious potential for distortion, the risk was even greater when preparing the specimen for the vacuum environment of the electron microscope. Since the primary constituent of cells is water, which will disperse as vapor once the air is evacuated, cells had to be subjected to treatments to remove all water without distorting the appearance of the cell structures being imaged. The most common way to do this was to displace the water with alcohol (or similar substance) and then remove the alcohol. However, even if no disruption is caused as the alcohol replaces the water, once the water is removed the different membrane-bound constituents shrink, and so change shape. One means of partially reducing this distortion was to place the specimen on a film before drying to preserve the shape in two dimensions. This, however, resulted in even greater distortion in the third dimension. Moreover, the chemical treatment itself involved transporting cell constituents to a foreign environment, which itself could distort the structures within the cell.

Finally, increasing contrast in a specimen required introducing a foreign substance of high atomic weight to bind with structures in the cell so as to increase their ability to deflect electrons. The substance that researchers initially seized upon was a fixative widely used in light microscopy, osmium tetroxide. While osmium yielded micrographs portraying structures, researchers did not know what osmium was reacting with and hence what they were seeing. Moreover, there was the question of how osmium should be applied; different techniques for its application (e.g., applications at different pH values) could result in different images.

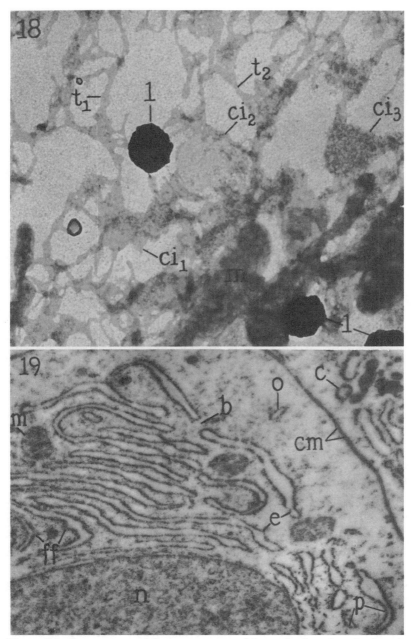

Figure 7.2. Comparison of the appearance of the endoplasmic reticulum in electron micrographs of glandular epithelia cells of the parotid of (a) a whole cell grown in tissue culture and (b) a thin section of (a) in situ. From Palade and Porter (1954), plate 62. Reproduced from *The Journal of Experimental Medicine* 100 (1954), plate 62, by copyright permission of the Rockefeller University Press.

Thus, a host of factors could render the structures in the micrographs into artifacts. They could be due to reactions with osmium or other electron stains. (Palade and Claude 1949 argued that the Golgi body was such an artifact.) The image that appeared could reveal real structure, but one greatly altered to produce a thin specimen or in the course of removing the water. Finally, even subjecting cells to the electron beam itself could "fry" the contents, leaving only artifacts.

Positron Emission Tomography

As with electron microscopy, the operation of a PET scanner relies on basic physical and chemical principles which were quite well understood when PET was employed as a technique for imaging activity in the brain in the 1980s. An unstable positron-emitting isotope, such as ^{15}O, is produced in a cyclotron by accelerating protons into the nuclei of oxygen atoms. To regain stability, the proton breaks down into two particles, a neutron and a positron. On breakdown, the neutron remains in the nucleus, while the positron travels away from the site of generation until it collides with an electron. At this point both the electron and the positron are annihilated, generating two gamma rays directed precisely 180° from each other. The PET scanner consists of a ring of gamma detectors, and an event is recorded when two gamma rays arrive simultaneously. From the points of arrival, the site of the positron annihilation can be ascertained through the techniques of computerized tomography. (Although not frequently discussed, possible risks of error exist due to the fact that detection of an annihilation depends upon the location of detectors. In addition, the distance the positron travels prior to annihilation can distort the spatial resolution of the data.)

To employ PET in brain imaging activity, a link between brain activity and positron emission and detection must be supplied. The first step is to inject the labeled oxygen into the blood so that the breakdown process described earlier occurs as the oxygen travels through the bloodstream. The brain activity of interest is neuronal firing, and it is assumed that blood flow will increase due to increased energetic demands when neuronal firing rates are increased. While a connection between metabolism and blood flow is very plausible, the mechanism linking neuronal firing and metabolism has not been established. There are several possible mechanisms for generating increased metabolism (neurotransmitter metabolism, action potential generation, etc.), some of which are not tightly linked to firing rates. Further, although it is natural to assume that increased activation reflects increased neural excitation, it is also possible that the increased blood flow is linked to inhibitory processing. Hence, while it seems plausible to interpret PET images as reflecting increased neuronal firing rates, it is possible that such interpretations are incorrect.[3]

While not denying these risks, I will concentrate my analysis on additional risks of artifacts that arise when one tries to relate what is taken to be increased neural activity recorded in the images generated by the PET scanner with cognitive activity. There are several sources of concern here. First, only relatively simple cognitive activities that can be performed repeatedly during the forty seconds of a scan can be studied, since the magnitude of any increased blood flow generated by the activity will be small and can only be detected, against the background of other activity and other noise, if repeated many times. That means that the cognitive activities will not be the ordinary ongoing cognitive activities of life, but specifically designed tasks (e.g., reading words from a list). Second, even in performing the highly simplified tasks that can be studied in the scanner, many different cognitive activities will occur, resulting in significant blood flow over a great deal of the brain. To create meaningful images, researchers try to focus on just one elementary cognitive operation or specific set of cognitive operations. This is done by imaging the person while performing two different tasks thought to differ, in that one employs the cognitive operations of interest while the other doesn't, and then subtracting the second image from the first (generating a difference image). For example, Petersen and associates (1989) subtracted the image produced which subjects read words aloud from the image produced when subjects read nouns, generated related verbs, and said those aloud. They thereby hoped to identify those brain areas required to generate the verbs (see Figure 7.3). This is known as the *subtractive method,* and it is recognized as one of the most controversial steps in the PET process.

The subtractive method was initially developed by F. C. Donders

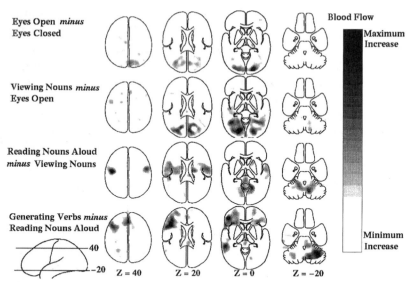

Figure 7.3. PET subtraction images of horizontal slices through the brain showing increased activations (a) with eyes open over eyes closed, (b) viewing nouns over eyes open, (c) reading nouns aloud over viewing nouns, and generating verbs over reading nouns aloud. The front of the brain is to the top; the Z values refer to millimeters above and below a horizontal plane through the brain marked "Z = 0." Figure is from Raichle, M. E. (1998). "Behind the Scenes of Functional Brain Imaging: A Historical and Physiological Perspective." *Proceedings of the National Academy of Sciences (USA)* 95: 765–772. Reprinted with permission of Marcus Raichle and the National Academy of Sciences.

(1868) for use in chronometric studies of cognitive process: reaction times for one task were subtracted from those for another task, and the difference was thought to reflect the time required for the additional processes required in the longer task. As Sternberg (1969) pointed out, the subtractive method assumed that the additional process was a pure insertion into a sequential set of processes, and this assumption might well be false. As a result, Sternberg advocated replacing the subtractive method in studies of mental chronometry by techniques intended to measure whether different tasks interfere with each other (see Posner 1978). Since PET is used to identify spatial areas active in one task that are not active in others, it *may* not suffer from the same limitations so long as one does not claim that the additionally activated areas are exclusively responsible for the additional operations. But one still must assume that the two tasks are identical except for the additional process invoked in one of them, an assumption that can always be challenged. Moreover, advocates of a strongly dynamical approach to cognition that views the performance of any cognitive task as involving the emergent interaction of brain components, and not the putting together of brain components involved in performing differential tasks, find it fundamentally suspect (van Orden 1997).

There are additional steps in the production of functional PET images at which artifacts could arise. First, to be interpretable, PET images must be mapped onto a common representation of the brain. The Tailerach Atlas (Tailerach and Tournoux 1988), which also provides stereotaxic coordinates for designating brain areas, has provided the standard reference. Not all brains, however, have the same shape, and individual brains have to be mapped onto the standard reference, which can easily induce distortions. Mapping onto a common reference frame is important for more than just interpreting PET images. The signal used in PET is itself rather weak, and usually researchers must average over several subjects in order to generate a statistically reliable signal. Thus, activations will show up only if they affect what are taken by the mapping scheme to be the same areas across multiple brains. Finally, the ultimate interest in PET is in determining what neuroanatomical structures are active in performing particular mental tasks, but for the most part the detailed neuroanatomy required to differentiate brain areas has been done only on nonhuman species such as the macaque. While there are many homologies between macaque and human brains, the precise topological relation of these areas in the two brains differs. (Deacon 1997 identifies a wide range of known deviations from the allometric projections from ape brains to humans.)

Thus, PET images are many steps removed from underlying brain activity, and there are many points where artifacts could arise. Especially important in this regard are the particular mental tasks for which imaging is done. Because of the constraints noted above, these tasks are often artificial tasks and hence may not indicate how the brain functions in ordinary mental activities. Moreover, task performance is likely to be altered by the requirements of the scanning situation, such as the requirement to repeat the same operation many times. (In fact, PET results themselves have revealed some of these alterations. After completing the PET study of subjects generating verbs in response to nouns to be discussed later, researchers decided to repeat the study with subjects who had practiced the task. Performance after even quite short practice periods resulted in different patterns of activation when the practiced lists were used. See Raichle et al. 1994.) The most critical factor is that the precise tasks employed vary from laboratory to laboratory. Not surprisingly, the activation patterns that are interpreted as resulting from the underlying cognitive processes vary, resulting in different laboratories' assigning what they claim to be the same mental operation to different (although usually adjacent) brain regions. As a result of these factors, there is a risk that the phenomenon of cognitive processing is substantially altered in ways that are not well understood (Stufflebeam and Bechtel 1997).

VISUAL PERCEPTION

Many of the same features identified in the previous section as arising with visualization instruments and potentially inducing artifacts also arise with ordinary human perception; thus attending to how we decide what are artifacts in perception can give us some clues as to how artifacts are detected and avoided in science. Phenomenologically, perception seems to be direct and immediate, concealing the fact that several steps intervene in generating our visual judgments. Illusory figures such as the one adapted from Kanizsa, with which I began, provide initial evidence that perceptual processes are less direct than they seem. The "new look" movement in visual perception, begun by Jerome Bruner and his colleagues in the 1940s, used stimuli such as tachistoscopically presented ambiguous playing cards (e.g., a red seven of spades), which subjects would see as normal cards (e.g., a red seven of hearts), to demonstrate the influence of cognitive factors (knowledge of the normal cards in a deck of playing cards) in perception.

Over the last forty years neuroscientists have discovered numerous brain areas mediating between impressions of light on the retina and perceptual judgments. The brain radically alters the input signal before a person arrives at a perceptual judgment. Starting in the retina, different classes of cells, identified as Pα and Pβ cells, respond to different features of the visual input. The Pα cells are relatively large and have radiating dendrites, large receptive fields, rapid conductance velocities, and transient responses. The Pβ cells, by contrast, are relatively small and have relatively small dendritic arbors, small receptive fields, and medium conductance velocities. As a result of these differences, the two classes of cells engage in different forms of information processing. The Pβ cells, because of their small receptive fields, are sensitive to high spatial frequencies, while the Pα cells, due to their transient firing patterns, are more sensitive to motion. This difference in processing is maintained in the lateral geniculate nucleus (LGN)–Pα cells project to the two magnocellular layers in the LGN, whereas Pβ cells project to the four parvocellular layers. As one progresses from the LGN to primary visual cortex (area V1), the divisions between processing systems become even more elaborate. There are three different processing streams through V1: the magnocellular stream primarily projects onto layers 4Cα and 4B of V1, while the parvocellular stream projects onto 4Cβ and 4A and then splits into two streams, one going through cells that stain as blobs with cytochrome oxidase and one through cells in the areas between these blobs (see Milner and Goodale 1995 for a review). After V1, processing spreads out even more. Felleman and van Essen (1991) have identified thirty-two different visual processing areas in the macaque; single-cell recording studies by Zeki and other researchers have identified different types of processing performed in different areas (Zeki 1993). At a large scale, the visual system splits into two different processing streams after V1, one progressing dorsally into the posterior parietal cortex while the other proceeds ventrally into temporal cortex. Reasoning from lesion studies in which deficits in the two streams generated very different error patterns, Miskin and Ungerleider (1982) proposed that the ventral stream was engaged in identifying objects (thus they called it a *what* pathway) while the dorsal stream was engaged in determining the locations of objects (thus they called it a *where* pathway). While it is now recognized that the actual organization is a good deal more complex (there are several points of information flow between the two streams), and that the distinction may not be between *what* and *where*, but between information needed for immediate action and information used for higher cognitive processing (Milner and Goodale 1995), this basic idea of two relatively independent streams processing different kinds of information about visual scenes has been generally supported. Thus, not only are there a large number of steps involved in generating our perceptual judgments, but the steps also transform the incoming visual information in a variety of ways reminiscent of the very ways in which electron microscopy and PET operate on the phenomena to generate images.

Why do illusions and information about the intermediate pathways not make us skeptics about our perceptual judgments? In part, no doubt, the reason why only a few philosophers have been tempted by such skepticism is that we have little choice when it comes to relying on our perceptual judgments. While it is possible to negotiate our world without relying on visual perception, it is notoriously difficult. But if forced to offer justification, we can offer a substantial number of arguments against skepticism. First is the power of the coherent image. Perception makes us aware of a world of objects which move, for the most part, in coherent ways. The coherence seen in the world of objects is especially noteworthy given the manner in which the perceptual system processes visual stimuli – separating, for example, the detection of motion, shape, and color from each other. The fact that from these separate channels of information processing we perceive a coherent world, rather than one in which shape, color, and motion are changing independently, makes it likely that each processing stream is responding to a common source and that the structure in our perception is a reflection of structure in the world we are perceiving. The coherence found in visual scenes is part of what makes it so hard to generate an illusion. Wertheimer (1912) was able to create the well-known phi illusion of moving lights by flashing individual lights on and off. But the parameters matter greatly in setting up such an illusion. The lights must go on and off in the right sequence and with the right timing. Otherwise, we simply see individual lights flashing. It gets increasingly more difficult to sustain such illusions if the perceiver is allowed to actively explore the environment (e.g., to move around the object).

If one continues to doubt the reality of what one sees, a second, and very powerful, way of overcoming doubt is to compare visual judgements with other sensory information. We can do this in two ways – by comparing our perceptual judgments either with judgments produced by our other sensory systems (touch, hearing, taste, etc.) or with the visual judgments of others. We detect that our visual system is being deceived when we unsuccessfully attempt to manipulate objects by touch, for example, on the basis of visual information. If you did test for yourself the two Kanizsa figures in Figure 1 by cutting out one and laying it on top of the other, then you relied on motor manipulation to check your initial visual perception (although in this instance the second sensory judgment was also based on vision). Trying to pick up an object generated by a hologram reveals the error in our perceptual judgement. Likewise, if your visual judgments do not agree with those of another person (perhaps someone with a different vantage point), you may wonder whether you have misperceived. However, in the vast majority of cases when we compare our visual judgments with those generated by our other senses or made by other agents, we confirm our initial judgments.

A final basis for rejecting visual illusions is reliance on what we know about the world. We look at an Escher drawing of a continuous downward flowing waterfall and realize that, since perpetual motion is impossible, something is wrong with the artist's portrayal of the water's movement.

One thing that we do not do in evaluating the reliability of visual perception is to appeal to theories about how our visual judgments are generated. As Hacking notes with regard to visual displays of instruments in the quotation at the beginning of the chapter, our visual judgments are robust under changes in theories as to how they are generated. The current account of visual processing reviewed here has only been developed in the last quarter century, and has not been the basis of any philosopher's defense of visual judgments against skepticism. These theories may be radically modified in the future, and such changes are extremely unlikely to have any impact on whether we trust visual perception. What they may do is give us a better appreciation of the conditions under which visual perception is likely to be unreliable. Beyond that, our capacity for reliable visual perception is a target for explanation, but is not dependent upon that explanation for justification.

EVALUATING THE USE OF NEW INSTRUMENTS IN BIOLOGY

The criteria scientists used to evaluate techniques such as electron microscopy and PET are much like those we use in evaluating visual perception. They did not invoke detailed theoretical understanding of the procedures for using the instrument. In attempting to improve techniques, scientists would use available theoretical knowledge (e.g., about how particles move in different media), but there were no comprehensive theories to justify procedures for slicing the cell, removing water, or staining with heavy metals. Similarly, PET researchers do not have detailed theories about how cognitive processes alter blood flow, how repeated performance of a task in the scanner alters cognitive processing, or what component cognitive tasks might be isolated by the subtractive method. (This is not to suggest that PET researchers are not extremely concerned with such questions. For example, investigators often appeal to the best theories in cognitive psychology to identify tasks. My point is simply that the success of PET in identifying brain areas responsible for these component processes is part of the evidence for the existence of the components themselves. This is why many researchers from more traditional cognitive science have eagerly embraced PET and other neuroimaging technologies.)

Evaluations of new scientific instruments are based on whether the techniques produce determinate results (sharply defined structures in the cell, well delineated areas of activation in the brain), results consilient with those generated by other techniques, and results that fit into emerging theories. One distinctive feature of such evaluations, though, must be kept in mind. New instruments are introduced so as to extend the ability to image biological phenomena. Thus, the search for consilience with instruments that measure exactly the same thing is futile, and researchers must settle for more indirect measures of consilience. However, in an important respect this is no different than seeking consilience among our different senses, since each gives different pieces of information and consilience stems from the way in which these different pieces of information are coordinated. In part, such coordination appeals to emerging theories about the phenomena themselves. The ability of images to provide data for plausible theories enhances the status of the images themselves. (Cf. Creath 1988, who proposes both consilience with other techniques and the ability to fit emerging theories as reasons for trusting perceptual judgements based on instruments, and also emphasizes how these figure in evaluation of our own perceptual judgments made without additional instruments.)

Electron Microscopy

The first attempt to apply electron microscopy to cells was made by Keith Porter in collaboration with an industrial electron microscopist, Ernest Fullam of Interchemical Corporation, in 1944 (Porter, Claude, and Fullam 1945). They made use of Porter's skill in tissue culturing and in fine manipulation of biological specimens to transfer a cultured cell from a fourteen-day-old chicken embryo to the grid of the electron microscope. In this first effort it was not known whether anything meaningful would be found or whether the micrographs would be totally uninterpretable. Upon looking at the images, though, Porter found structure. Only the fringes of the cell were visible, but there one could see interconnected strands and vesicles of small dimensions and relatively low density, which Porter characterized as a "lace-like reticulum." But did this structure correspond to a part of a live cell? For Porter, the appearance of such a reticulum accorded with the popular hypothesis that cell structure was maintained by a cytoskeleton (Needham 1942). Thus, even in interpreting his first micrograph, Porter's acceptance of the structure as real was influenced by his interpretation of it.

The lacelike reticulum was not the only cytoplasmic structure revealed in the early electron micrographs. Another was the mitochondrion. Although the mitochondrion had been identified in light microscopy, its internal structure had not. As researchers developed tools for thinly slicing cells and varying the application of osmium tetroxide as a fixative, they began to discover structure in the mitochondrion. Palade (1952a) in particular identified what he took to be regular infoldings in the inner mitochondrial membrane, which he called *cristae*. Sjöstrand (1956b) took issue with Palade's interpretation, arguing that these internal structures only appeared as infoldings of the inner membrane as a result of postmortem swelling that resulted from not fixing the cell promptly. Palade's interpretation, however, was more generally accepted, partly because his detailed study of fixation procedures with osmium tetroxide supported belief that his micrographs were more reliable. But in fact, Palade's evaluation of fixation procedures itself relied in part on whether they provided the most detailed and sharp micrographs (and avoided obvious signs of artifact), so his evaluation of fixation procedures did not have a truly independent basis.

Another factor supporting Palade's interpretation was that, partly as a result of the development of yet another new research technique, cell fractionation, a function was readily assignable to the infolded membrane. Fractionating cells in the ultracentrifuge isolated a cell component identified as mitochondrial in origin (Claude 1943) that contained several of the critical enzymes for cellular respiration (Hogeboom, Schneider, and Palade 1948). Independently, biochemists had concluded that respiration required membranes (the reason for this was not discovered until the 1960s). Palade's identification of cristae thus fit a cohesive account that incorporated consilient evidence from cell fractionation and biochemistry (Bechtel 1990).

Let us return now to the lacelike reticulum Porter identified in his first micrograph. After his initial foray, Porter continued to develop the techniques of electron microscopy of tissue-cultured cells and identified a similar structure in a variety of other cells. Porter and Kallman (1952) named the structure the "endoplasmic reticulum," since the appearance suggested a reticular structure and was generally found only in the inner region or endoplasm. Researchers producing electron micrographs from tissue slices also noticed a structure in these areas of the cell but saw it not as a vascular structure but as pairs of membranes (Sjöstrand 1953) or filaments (Dalton 1953). The reason for the difference is straightforward: working with slices of cells, researchers only saw the vesicles when they lay in the plane of the slice; otherwise they would appear, if at all, as vacuoles.

Porter's claims were widely discounted, largely because they were based on tissue-cultured cells. In an attempt to show that the endoplasmic reticulum was the source for the paired membranes or filaments, Palade and Porter (1954) carried out a comparative study of thin-sliced preparations and both whole tissue-cultured cells and thin-sliced preparations from tissue-cultured cells (see Figure 2). They then appealed to the fact that "[t]he same 'broken down' appearance is encountered . . . in cultured cells when examined in sections" as "good evidence that the appearance mentioned is the result of sectioning" (p. 647).

This attempt to establish consilience between the techniques of thin slicing and tissue culturing cells was not alone sufficient to convince the skeptics (see Sjöstrand 1956a). But by the late 1950s, with the development of a more comprehensive perspective, skepticism as to the reticular structure began to diminish. Both Palade (1953) and Porter (1954), working with thin sections, identified particles on substantial portions of what they took to be pieces of the endoplasmic reticulum. Turning to biochemical tools, Palade and Siekevitz (1956) established high levels of RNA in these particles, which came to be known as ribosomes, and proposed that they played a role in protein synthesis. This required some means for moving RNA from the nucleus to the ribosomes. Porter then pointed to the apparent connections between the smooth portions of the endoplasmic reticulum and pores in the nuclear membrane and offered the speculative proposal that the endoplasmic reticulum "through its cytoplasmic patterns and pathway serves as a vehicle for the transfer of genetic information to the cytosome" (Porter 1961: 654). Although this proposed function is different than in Porter's initial suggestion of a cell cytoskeleton, the discovery of a function for the endoplasmic reticulum secured it a place in an emerging theoretical model.

From this brief review, we can identify how the factors noted at the beginning of this section figured in establishing the credibility of electron microscopy. First, the details of the images were themselves extremely compelling; and as techniques were refined, even more detail appeared, further increasing the convincing power of the micrographs. Further, consilience between alternative models of preparation (tissue culturing and cell fractionation) and between electron microscopy itself and other techniques (cell fractionation) further enhanced credibility. Finally, at several stages theories of cell function that were themselves supported by evidence from electron microscopy reciprocally enhanced the credibility of the electron microscopy itself.

Positron Emission Tomography

As with electron microscopy, there was a real potential that PET would fail to produce any intelligible images. In most PET imaging a threshold is set that the summed average of activation over multiple subjects must exceed in order for the increased activation in an area to count as statistically significant. Given that multiple performances of a particular task by several individuals are required to generate an image, it would seem entirely likely that individual performances would vary and cancel each other out, or yield a pattern in which the pixels whose activations are statistically significant would be distributed randomly over cortex. The fact that continuous areas are all activated above threshold suggests that there must be something giving rise to this image; that is, that there is some structure in the brain that the PET image is reflecting. For those who are not actually involved in creating PET images but simply see them in published reports or in talks, the multiply colored images (which are themselves just a way of representing numerical data) are often transfixing, and many audiences assume they are definitive data and do not question whether they might be artifacts.

How do PET researchers try to show that these images are not artifacts? One strategy is to demonstrate that the results are consistent with the results of other ways of studying cognitive function in the brain, such as lesion studies and single cell recording. Prior to the advent of PET, these tools had provided a modestly rich account of the tasks performed by different areas in primate cortex. (For the most part, these studies were not performed on humans, so to establish consilience researchers were required to project across brains from different species.) But these studies had provided good grounds for believing that most visual processing occurs in occipital cortex and surrounding areas of temporal and parietal cortex. Thus, when in the first step of their study of single word processing, Petersen and his colleagues (1989) simply required subjects to passively view individual words and generated activations in areas that were projected to lie in and around occipital cortex, they established a degree of consistency between PET and these other techniques. The PET results further indicated that a word-form identification area was located in prestriate cortex, a claim that was consilient with other PET studies (Petersen et al. 1990) showing activation in these areas when words and pseudowords were presented, but not when random letter strings or false fonts not comprised of English letters were presented. They found further support in neuropsychological findings that lesions in these areas resulted in pure alexia, or the inability to read words (Damasio and Damasio 1983).

In the Petersen and colleagues (1989) study, the images from these tasks were subtracted from images generated when subjects were required to pronounce the nouns they had either seen or heard. The resulting subtraction image yielded bilateral activations in the motor and sensory face areas as well as in the cerebellum. They found these results also to be highly credible, since they fit other evidence about areas involved in motor, especially articulatory processing (for example, that lesions in the most focal parts of Broca's area are known to result in speech production deficits).

The critical new task in the Petersen study required subjects to generate verbs corresponding to nouns they either heard or saw. This task was conceived as adding a semantic component to the pronunciation task. It was expected that the areas of increased activation, after subtracting the areas active in the corresponding noun pronunciation task, would be those involved in semantic processing. The researchers obtained a complex pattern of results, the most surprising feature of which was increased activation in the left prefrontal cortex and no increased activation in Wernicke's area, the area traditionally associated with semantic processing (due to the kinds of deficits accruing with lesions in that area). Accordingly, the authors proposed that it is the left prefrontal cortex, not Wernicke's area, that is the locus of semantic processing.

Since their result was contrary to a long history of lesion studies, the burden was the Petersen researchers to show that their results were not artifacts. One can discount the lesion data in part by noting that lesions only indicate weak points in a processing system (through which information is conducted, for example), not necessarily the location where a failed process is performed. But Petersen and his associates required a more positive defense for their alternative proposal. One strategy they employed was to emphasize consilience with earlier, non-PET blood flow studies. However, since PET also relies on blood flow, that raises the possibility that both sets of studies rest on artifacts due to blood flow.

To make credible the claim that the apparent increased activation in left anterior inferior prefrontal cortex is due to semantic processing, they needed other evidence that this area is involved in semantic tasks, and theoretical accounts of what role it might play. There was some support on both fronts. First, there was other evidence indicating a role for left prefrontal cortex in semantic tasks. The authors themselves reported PET scans on five subjects carrying out a semantic judgment task (did words refer to objects in the same category?) which indicated activation in a very similar area of prefrontal cortex (although these activations did not reach the threshold for statistical significance). Further, in a follow-up study Frith and his colleagues (1991) found activation in a semantic task in both Wernicke's area and prefrontal areas. The Frith team presented their results as *disconfirmation* of Petersen's results, since they did get activation in Wernicke's area, and they attributed the prefrontal activation to "intrinsic generation rather than semantics" (p. 1146). Yet, since all of the input conditions in the Frith study involved auditory input, Petersen and Fiez (1993) could argue that the activations that Frith and associates found in Wernicke's area had to do with auditory processing, not semantic processing. This left only the activation in prefrontal cortex as a candidate for semantic processing. Second, the authors noted the work of Goldman-Rakic (1987) showing that lesions in anterior prefrontal cortex in monkeys leave the animals unable to withhold responses to stimuli, a cognitive process that is likely to be important in semantic processing (see also Deacon 1997, who draws the linkages between the capacity to withhold responses and semantic processing).

Finally, the Petersen study advanced a flowchart model of how the various areas identified in the different tasks on which they scanned subjects might interact in the processing of single words (Figure 7.4). One important feature of the model is that it identifies two routes leading to speech output, one directly from visual word-level encoding to motor processing and another through semantic

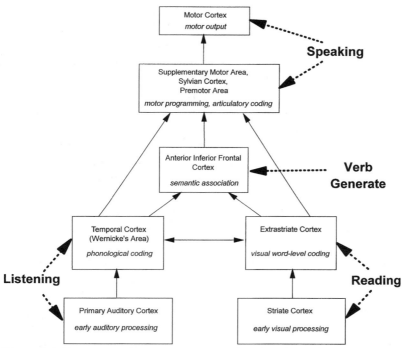

Figure 7.4. Two routes for processing visual and auditory language inputs (after Petersen et al. 1989). Note that neither the direct route from visual processing to motor output nor that through semantic association employs Wernicke's area.

processing. In neither pathway is Wernicke's area involved. Petersen and colleagues claimed independent support for this model from research on reading aloud, where dual-route models of reading originated. In these models one route is construed as relying on semantic processing (used to read words with nonstandard pronunciations such as *pint*) while the other bypasses semantic processing and instead employs grapheme-to-phoneme correspondence rules (to permit reading nonwords such as *rint*). Dual-route models have received independent support from neuropsychological research that reveals a double-dissociation between patients who can read words with nonstandard pronunciations but not nonwords, and patients who can read nonwords but not words with nonstandard pronunciations (Coltheart 1985), but they have also been challenged by a variety of investigators (Plaut 1995). By drawing upon and expanding the dual-route framework so as to distinguish phonological coding and visual encoding of inputs, Petersen and his colleagues challenged the claim that all processing of words (whether presented auditorily or visually) involved Wernicke's area. They thereby also embedded their results with the new imaging technique in a plausible theoretical framework.

As with the electron microscope, researchers invoking PET have defended the reliability of the technique by emphasizing both its consilience with other techniques and the theoretical plausibility of the data generated by it. I have focused on one piece of data generated by a relatively early PET study which conflicted with more standard views and tried to show that even here the researchers have defended the result by seeking consilient evidence as well as advocating a theoretical framework in which the result made sense (for further discussion, see Stufflebeam and Bechtel).

CONCLUSIONS

While seeing does not always justify believing, generally it does. The same is true with established imaging techniques in science. But early in the introduction of new imaging techniques, researchers are often greatly concerned as to whether their techniques are generating artifacts. In this chapter I have focused on two imaging technologies which have come to play important roles in biology – electron microscopy as a means of gaining evidence about cell structure, and PET as a means of identifying functional components in the brain. I have tried to show that the reasons for doubting the reliability of new techniques have parallels in the case of ordinary perception. The ways in which we vindicate perception and reject skepticism, however, also find parallels in the ways in which scientists evaluate their techniques and overcome doubts about artifacts. The detail found in the images is itself compelling evidence that scientists are not dealing with artifacts but with images of real structure or function. But beyond that, they rely on consilience with other techniques, sometimes already established ones, often ones just in the process of being established. Finally, the ability of the techniques to support theoretical models reciprocally is often taken as evidence that the techniques are revealing real structure or function and not artifacts.

NOTES

1. Support for research on the history of electron microscopy was provided by the National Endowment for the Humanities (RH-21013-91), which is gratefully acknowledged.
2. Although I am not focusing on light microscopy in this chapter, it should be noted that it raised the same range of epistemic issues. For example, biological materials typically had to be fixed and stained, and both steps involved chemical alteration of the contents of the cell. One of the most useful stains for neurobiology was Golgi's silver stain; it was useful in large part because it stained only some of the neurons in a preparation, and thus revealed their individual identities. It is still not understood why it behaves in this way.
3. I thank Susan M. Fitzpatrick for pointing out these concerns, as well as those noted at the end of the previous paragraph.

REFERENCES

Bechtel, W. 1990. "Scientific Evidence: Creating and Evaluating Scientific Instruments and Research Techniques." *PSA 1990* 1: 559–572.

Bechtel, W. 1993. "Integrating Sciences by Creating New Disciplines: The Case of Cell Biology." *Biology and Philosophy* 8: 277–299.

Bruner, J. S. and L. Postman. 1949. "On the Perception of Incongruity: A Paradigm." *Journal of Personality* 18: 206–223.

Claude, A. 1943. "Distribution of Nucleic Acids in the Cell and the Morphological Constitution of Cytoplasm." In Jacques Cattell, ed., *Biological Symposium. Volume X. Frontiers of Cytochemistry,* 111–129. Lancaster, PA: Jacques Cattell Press.

Coltheart, M. 1985. "Cognitive Neuropsychology and the Study of Reading." In M. I. Posner and O. S. M. Marin, eds., *Attention and Performance X.* Hillsdale, NJ: Lawrence Erlbaum.

Creath, R. 1988. "The Pragmatics of Observation." *PSA 1988* 1: 149–153.

Dalton, A. J. 1953. "Electron Microscopy of Tissue Sections." *International Review of Cytology* 2: 403–417.

Damasio, A. R. and H. Damasio. 1983. "The Anatomic Basis of Pure Alexia." *Neurology* 33: 1573–1583.

Deacon, W. 1997. *The Symbolic Species: The Co-evolution of Language and Brain.* New York: Norton.

Donders, F. C. [1868] 1969. "Over de snelheid van psychische processen. Onderzoekingen gedaan in het Psyiologish Laboratorium der Utrechtsche

Hoogeschool: 1868–1869." Tweede Reeks, II, 92–120. Translated by W. G. Koster as "On the Speed of Mental Processes." *Acta Psychologica* 30: 412–431.

Felleman, D. J. and D. C. van Essen. 1991. "Distributed Hierarchical Processing in the Primate Cortex." *Cerebral Cortex* 1: 1–47.

Frith, C. D., K. J. Friston, P. F. Liddel, and R. S. J. Frackowiak. 1991. "A PET Study of Word Finding." *Neuropsychologia* 12: 1137–1148.

Hogeboom, G. H., W. C. Schneider, and G. E. Palade. 1948. "Cytochemical Studies of Mammalian Tissues. I. Isolation of Intact Mitochondria from Rat Liver: Some Biochemical Properties of Mitochondria and Submicroscopic Particulate Material." *Journal of Biological Chemistry* 165: 619–635.

Latour, B. 1987. *Science in Action: How to Follow Scientists and Engineers Through Society.* Cambridge: MIT Press.

Milner, A. D. and M. A. Goodale. 1995. *The Visual Brain in Action.* Oxford: Oxford University Press.

Needham, J. 1942. *Biochemistry and Morphogenesis.* London: Cambridge University Press.

Palade, G. E. 1952. "A Study of Fixation for Electron Microscopy." *Journal of Experimental Medicine* 9: 285–297.

Palade, G. E. 1953. "A Small Particulate Component of Cytoplasm." *Journal of Applied Physics* 24: 1419.

Palade, G. E. and A. Claude. 1949. "The Nature of the Golgi Apparatus. II. Identification of the Golgi Apparatus with a Complex of Myelin Figures." *Journal of Morphology* 85: 71–111.

Palade, G. E. and Porter, K. R. 1954. "Studies on Endoplasmic Reticulum." *Journal of Experimental Medicine* 100: 641–655.

Palade, G. E. and P. Siekevitz. 1955. "Liver Microsomes: An Integrated Morphological and Biochemical Study." *Journal of Biophysical and Biochemical Cytology* 2: 346–375.

Petersen, S. E. and J. A. Fiez. 1993. "The Processing of Single Words Studied with Positron Emission Tomography." *Annual Review of Neuroscience* 16: 509–530.

Petersen, S. E., P. T. Fox, M. I. Posner, M. Mintun, and M. E. Raichle. 1989. "Positron Emission Tomographic Studies of the Processing of Single Words." *Journal of Cognitive Neuroscience* 1: 153–170.

Petersen, S. E., P. T. Fox, A. Snyder, and M. E. Raichle. 1990. "Activation of Extrastriate and Frontal Cortical Areas by Visual Words and Word-like Stimuli." *Science* 249: 1041–44.

Plaut, D. C. 1995. "Doubled Dissociation without Modularity: Evidence from Connectionist Neuropsychology." *Journal of Clinical and Experimental Neuropsychology* 17: 291–321.

Porter, K. R. 1954. "Electron Microscopy of Basophilic Components of Cytoplasm." *Journal of Histochemistry and Cytochemistry* 2: 346–375.

Porter, K. R. 1961. "The Ground Substance: Observations from the Electron

Microscope." In J. Brachet and A. E. Mirsky, eds., *The Cell: Biochemistry, Physiology, Morphology.* New York: Academic.

Porter, K. R., A. Claude, and E. Fullam. 1945. "A Study of Tissue Culture Cells by Electron Microscopy." *Journal of Experimental Medicine* 81: 233–246.

Porter, K. R. and F. L. Kallman. 1952. "Significance of Cell Particulates as Seen by Electron Microscopy." *Annals of the New York Academy of Science* 52: 882–891.

Posner, M. I. 1978. *Chronometric Explorations of Mind.* Hillsdale, NJ: Erlbaum.

Raichle, M. E., J. A. Fiez, T. O. Videen, A.-M. K. MacLeod, J. V. Pardo, P. T. Fox, and S. E. Petersen. 1994. "Practice-related Changes in Human Brain Functional Anatomy during Nonmotor Learning." *Cerebral Cortex* 4: 8–26.

Sjöstrand, F. S. 1953. "Electron Microscopy of Mitochondria and Cytoplasmic Double Membranes." *Nature* 171: 30–32.

Sjöstrand, F. S. 1956a. "Electron Microscopy of Cells and Tissues." In G. Oster and A. W. Pollister, eds., *Physical Techniques in Biological Research,* 241–98. New York: Academic.

Sjöstrand, F. S. 1956b. "The Ultrastructure of Cells as Revealed by Electron Microscopy." *International Review of Cytology* 5: 455–533.

Sternberg, S. 1969. "The Discovery of Processing Stages: Extension of Donders' Method." *Acta Psychologica* 30: 276–315.

Stufflebeam, R. S. and W. Bechtel. 1997. "PET: Exploring the Myth and the Method." *Philosophy of Science* 63: supplement.

Talairach, J. and P. Tournoux. 1988. *Co-planar Stereotaxic Atlas of the Human Brain.* New York: Thieme.

Van Orden, G. and K. R. Papp. 1997. "Functional Neuroimages Fail to Discover Pieces of Mind in the Parts of the Brain." *Philosophy of Science* 63: supplement.

Wertheimer, M. 1912. "Untersuchungen über das Sehen von Bewegung." *Zeitschrift für Psychologie* 61: 161–265.

Zeki, S. M. 1993. *A Vision of the Brain.* Oxford: Blackwell Scientific.

Part Three

The Nature and Role of Argument

Chapter 8

The Logic of Discovery in the Experimental Life Sciences

FREDERIC L. HOLMES

The subject of experimentation in science has been raised from its purported subordination to theory in the philosophy of science, and from the marginal position to which it is alleged to have been relegated by an earlier history of science centered on ideas, to imposing heights of interest in both fields. Sociologists and ethnomethodologists have further contributed to a growing preoccupation with experimentation by subjecting laboratory practice to the scrutiny of their special investigative frameworks. Where experiments were once viewed primarily as means to test theories, they are now repeatedly proclaimed to have a "life of their own."

A plethora of new approaches to experimentation has appeared. Bruno Latour and Steve Woolgar have treated the life of a laboratory as that of a strange culture that they have attempted to interpret without benefit of deep knowledge of the science pursued within it.[1] Steven Shapin and Simon Schaffer have described the experimental reports of Robert Boyle as a "literary technology" designed to "compel assent" to "matters of fact" obtained through experimental procedures carried out with an air pump.[2] Ian Hacking has viewed experimental intervention as the prime guarantor of the reality of unobservable scientific entities such as electrons.[3] Hans-Jörg Rheinberger has shifted focus from the individual experiment to the experimental system, which acquires a momentum that often leads the experimentalist in directions that could not have been foreseen.[4] Robert Crease has offered the argument that experimentation can be more fully understood by exploring its analogy to theatrical performance.[5] Herbert Simon has applied computer programs to recapitulate experimental discoveries and to understand them according to general problem-solving heuristics.[6]

Whereas earlier discussions of the role of experimentation most often drew their examples from the physical sciences, and more particularly from experimental physics, the newer studies have included prominently cases from the experimental life sciences. Kenneth Schaffner has focused most explicitly on recent work in the biomedical sciences to illustrate issues in the philosophy of science that include the nature of experimentation.[7] A compelling reason for the growing interest in biological experimentation is the dominance of that activity in the science of the past several decades. Biological experimentation is, however, not of recent origin. The earliest recorded scientific experiment in the Western tradition, contained in the papyrus entitled *Anonymous Londenensis,* states that the quantity of "invisible" transpiration can be determined by measuring the difference between the weight of the food ingested by a bird and that of its excretions. The last great physician of classical antiquity, Galen, regularly resorted to experimental interventions to settle disputed questions about physiological functions. In the early modern period, Vesalius performed experiments as well as dissections, and William Harvey's discovery of the circulation was brilliantly supported by experimental arguments. Through the remainder of the seventeenth and the eighteenth centuries individual investigators continued to perform significant experiments that defined critical functions such as whether digestion was a chemical or mechanical process, until, in the early nineteenth century, physiology became established as the first biological science dependent on systematic, institutionalized experimental investigation. Consequently, the nature of biological experimentation cannot be treated only as an extension of methods founded in the physical sciences. The life sciences must be seen instead as a formative site for the origin and development of experimentation in general.

Although each of the characterizations of experimentation mentioned here draws either on case histories or on contemporary case studies, they have been framed by the epistemological outlooks of philosophy, sociology, anthropology, and the cognitive sciences. How should historians of the experimental sciences respond to the multiple interpretative possibilities emanating from these neighboring disciplines? Too easily, I believe, we have been drawn into alliances which tempt us to use historical examples to validate a framework that appears imposingly argued, or one that seems to embody new questions and new directions that we avidly embrace in a perpetual quest for novel perspectives. Historians of science need to develop a balance between openness to ideas from our neighbors and the self-assurance that preserves autonomy. We should be able to entertain philosophical, sociological, anthropological, and cognitive explanatory schemata, to measure their fit with the historical events we study, and occasionally to suggest further articulations of such schemata that may improve their fit, but retain the independence of our historical investigative canons. Sometimes we may be in a position to adjudicate between competing general accounts of experimentation, but we need not choose to work within the confines of any one or two of them. The simultaneous presence of several disciplines offering us explanatory schemes can be, if we are strong enough to engage with each of them and to avoid seduction by any one of them, a source both of interpretative richness and of our own disciplinary identity.

My own subject of historical investigation for more than two decades has been what I sometimes call the "fine structure" of experimental investigation. Choosing case studies for which a dense documentary record of laboratory operations has survived, I have sought to reconstruct individual research trajectories at the level of daily thought and action. The cases I have chosen have come mainly from the experimental life sciences, though some have been at the intersection of chemistry and the chemistry of life. I have been generally aware of the recent sociological, philosophical, and cognitive approaches to experimentation, and have sometimes commented on their relation to my cases, but have also sought a framework that expresses my personal experience as a specialist immersed in a particular genre of documents. The concept, or metaphor, of the investigative pathway has helped me to define both my overall "unit" of historical analysis and the shape of sustained individual experimental activity. It is time, however, for me to confront more systematically the views of philosophers, sociologists, cognitive scientists, and others who examine experimentation from different disciplinary perspectives, in the expectation that dialogue between those for

whom history provides illustrations and evidence for broader generalizations, and those for whom historical cases are the main subject,
will be mutually enlightening.

As a first step toward such interactions, I shall sketch in this
chapter some outlines for further discussions of an issue in which
philosophers have long been interested: whether scientists can reason their way to discoveries, or whether only the context of justification is logical. My ability to make a contribution to a topic that has
already been thoroughly examined rests on the degree to which the
types of evidence with which I have worked, and the intimate level
of resolution at which I have attempted to follow experimental activity, can further illuminate questions customarily discussed at a
broader level.

The main obstacle to a philosophical account of discovery during
recent decades has been the opinion maintained by Karl Popper that
"every discovery contains 'an irrational element', or 'a creative intuition', in Bergson's sense." Quoting Einstein to the effect that there is
"no logical path" toward the discovery of universal laws, Popper
asserted that the "act of conceiving or inventing a theory seems to me
neither to call for logical analysis nor to be susceptible to it." If, as he
believed, "the work of the scientist consists in putting forward and
testing theories," then the work of the philosopher consists in an
analysis of the logic of testing theories.[8]

Beginning with Norwood Russell Hanson in the 1950s, other philosophers have challenged Popper's position. By the late 1970s the
"friends of discovery" had acquired sufficient momentum to produce two collective volumes of papers dealing with the "logic and
rationality" of scientific discovery. In an introductory paper, Thomas
Nickles summarized the case that the context of discovery cannot be
separated from the context of justification. Philosophers must either
confront discovery, or ignore "not only the most interesting phases of
scientific research but also . . . phases highly relevant to epistemology, e.g., to the theory of rationality and the understanding of conceptual change and progress in science." Acknowledging that there
does not exist a deep logic of discovery independent of specific subjects of inquiry, Nickles maintained that "less rigorous rules, rou-
tines, [and] . . . heuristics for discovery" do exist, and called for
philosophers to

join historians and other students of science in the descriptive task of making intelligible to reason (insofar as possible) actual cases of creative discoveries. The epistemologists' aims may not be identical with the historians', but we are engaged in a common task of providing accounts of the discovery process.[9]

That this call to collaboration has so far not yielded a large response is probably due not to the reticence of philosophers to engage the subject, but to the fact that the majority of historians have been less concerned with the intellectual processes of discovery than with the social aspects of scientific practice.

"The thesis that the process of scientific discovery involves logically analyzable processes," Kenneth Schaffner wrote in 1992, "has generally not been a popular one in this century." Schaffner's defense of a logic of discovery is especially pertinent to the concerns of the present volume, because he discusses discovery in the biomedical sciences rather than in the more commonly discussed physical sciences. With a few exceptions, Schaffner contends, theories in biology and medicine are not "universal," but "middle range." That is, they do not necessarily apply to all organisms. Many of them can be characterized as a "series of overlapping interlevel temporal models."[10] This characteristic of biological theories immediately suggests that they may not be subject to the rejection of a logic of discovery that Popper justified on the grounds that "inference to theories, from singular statements which are 'verified by experience' . . . is logically inadmissible." For Popper, theories were "universal statements," and one can never reason logically from singular statements to universal statements because the "difficulties of inductive logic . . . are insurmountable."[11]

Schaffner did not invoke the middle range nature of biological theories as a strategy to evade Popper's position, because he was more concerned with the later objections of Carl Hempel to a logic of discovery. Differentiating within the context of discovery between a "logic of generation" and one of "preliminary evaluation," Schaffner provided a series of arguments for believing that both phases can be largely, if not entirely, rational processes. Discoveries emerge from the efforts to solve scientific problems within a "domain" delineated by previous practice. These problems "can evoke a sufficient set of constraints to enable them to serve as the source of regulated scientific inquiry." As a means to understand the nature of this activity he discussed, among other things, Herbert Simon's problem-solving heuristics.

Schaffner illustrated and developed his position through the historical example of the discovery of the clonal selection theory of antibody formation by Macfarlane Burnet. From Burnet's autobiographical account of this discovery and his publications on the subject, and the background knowledge comprised mainly of Niels Jerne's earlier natural selection theory, Schaffner was able to offer a partial rational reconstruction of the generative phase of Burnet's discovery and a fuller account of the phase of preliminary evaluation. Even though an "irrational 'eureka' element" and other gaps might remain in the reconstructed generative phase, Schaffner asserted that it was possible to "make some progress in understanding the discovery process."[12]

Burnet had "given future investigators enough clues to at least partially mimic the private generative process," Schaffner wrote, but "it would clearly be utopian to expect, at this point in the development of a logic of generation, that Burnet's discovery of the clonal selection theory could be fully rationally reconstructed."[13] We might ask, however, whether it is the lack of a more fully developed philosophical "logic of generation" or a more complete historical record that is the barrier here to a full rational reconstruction of the discovery. Besides the public presentations of the clonal selection theory, Schaffner had only Burnet's succinct retrospective memory of the events leading to the discovery to guide his reconstruction. Might a full reconstruction be possible in cases where a daily record of the investigative pathway leading to a discovery has been preserved?

Two discoveries I have studied which fulfil the latter criterion are those by Hans Krebs of the ornithine cycle of urea synthesis and the citric acid cycle, now widely known as the Krebs cycle. The laboratory notebooks kept by Krebs contain a nearly uninterrupted record of his experimental pathway from the time he entered the laboratory of Otto Warburg in 1926 until after the publication of the citric acid cycle in 1937. In the case of the ornithine cycle, which he announced in the spring of 1932, the recovery of the notebook of his student Kurt Henseleit, together with his own notebooks, enables a reconstruction of every experiment he performed during the nine months between the time he took up the problem of urea synthesis and the time that he published his solution to the problem. The case of the citric acid cycle is more complex, because it represents the outcome of several years of intense but intermittent efforts to find the pathway of oxidative carbohydrate metabolism. The record is not quite complete, because the notebook of his assistant during the year preceding the discovery is not extant. A copy of a doctoral dissertation in which William Arthur Johnson pencilled in the dates of each of the experiments described there enables us, however, to fill in most of the gaps left by the record of the experiments that Krebs himself carried out over the years leading to the discovery.

The ornithine and the citric acid cycles are examples of "middle-range" theories that, according to Schaffner, include "most theories propounded in biology and medicine." They do not apply to all "organisms available on earth,"[14] but are widespread among animals. There are, moreover, families of similar models which account for the corresponding functions in organisms in which these specific cycles do not occur. The shape of the inquiries that led Krebs to both discoveries was tightly constrained by prior knowledge of metabolic reactions and rules for proposing and testing candidate reactions. Nevertheless, neither discovery was anticipated in detail by Krebs or by contemporary biochemists, and both are counted as "great" discoveries in the field. They cannot, therefore, be discounted as lacking significance or creativity.

Elsewhere I have reconstructed the entire research pathway of Hans Krebs from 1926 to 1937 in fine detail, reaching, for the more critical stages of his investigations, to the level of each individual experiment performed. Because Krebs's methods enabled him to carry out a set of from six to more than a dozen experiments each morning and another set in the afternoon, the reconstruction is very long, filling most of two volumes in which I have followed the first half of Krebs's scientific life.[15] It is not possible to recapitulate in this essay the stretches of that research pathway relevant to the ornithine and citric acid cycles. I can give only brief summaries of the several stages that marked his pathway to each of these two discoveries, and then discuss some of the features that one can extract from the full reconstructions.

For both investigations Krebs employed manometric tissue slice methods that he had learned in the laboratory of Warburg. These methods enabled him to measure the respiratory exchanges of intact cells surviving in slices or minces of tissue maintained in a solution whose composition was similar to that of the fluid medium surrounding the tissues in the whole organism. By adapting classical methods for the quantitative analysis of organic compounds to the minute quantities consumed or produced in such tissue slices, Krebs was able to measure also the rates at which a suspected metabolite disappeared or appeared. A rule generally accepted in the field was that to be considered an intermediate in a reaction pathway the substance in question must be transformed at a rate at least as rapid as that of the overall reaction in which it was suspected to take part.

When Krebs took up the problem of urea synthesis, in July 1931, his first experimental step was to adapt to his manometric methods an existing procedure for measuring the rate of formation of urea by means of the CO_2 it produced through the action of the enzyme urease. Then he began to test currently held theories of urea formation, such as that the urea nitrogen is derived from the deamination of amino acids, that NH_4 is an obligatory intermediate, and that pyrimidines might provide a separate source of urea resulting from nucleic acid metabolism. Soon he turned over the main experimental work to his first student, Henseleit, who systematically tested various conditions on the rate and reliability of the formation of urea. During his first two months of research on the question, Krebs established that his tissue slice method provided a precise, reliable way to study synthetic metabolic pathways such as the formation of urea; but he did not find out anything about the process that was not already known or suspected.

During October, Krebs and Henseleit began testing the effects on the rate of urea formation in liver tissue of various substances, including several amino acids, that might serve as sources of the urea nitrogen. Krebs compared the resulting rates in each case with that caused by the addition of ammonia. The experiments of this period imply that Krebs either assumed, or was testing, the generally accepted view that ammonia is an intermediate. Late in the month Henseleit tested the amino acid alanine alone and together with NH_4, an experiment connected with a "guess" by Krebs that the amino acid might contribute one of the urea nitrogens directly, with ammonia providing the other nitrogen. None of these experiments yielded decisive results. Only arginine produced much larger quantities of urea than ammonia alone, an expected effect, because arginine was known to give rise to urea through the enzyme arginase contained in the liver.

During the first half of November, Krebs seemed to turn away from these unsuccessful efforts to identify the source of urea nitrogen to explore instead the effects of substances, such as sugar and intermediates of carbohydrate metabolism, that might exert more general effects on the formation of urea by stimulating the overall metabolism of the cells. On November 15, however, he attached to one such set of experiments two runs testing the effects, respectively, of ornithine alone and of ornithine in combination with ammonia. The latter yielded a dramatic increase over the rate due to ammonia alone.

This result, unexpected because Krebs had no reason to think that one amino acid would differ from another in its effects on the formation of urea, led him to begin a search to define the nature of the effect. To find out whether the effect was unique, or only one out of a class of similar effects, he tested other compounds analogous in composition to ornithine. None of the other substances he tried produced, either alone or in combination with ammonia, a comparable increase.

By early in 1932, Krebs was concentrating more and more of his attention on the nature of the ornithine effect itself. The titles and designs of some of his experiments imply that he entertained the possibility that ornithine was a nitrogen donor in the reaction; others imply that he thought it only "influenced" the formation of urea from ammonia. As accumulating evidence gradually ruled out the first alternative, the second came to the fore, meaning that the effect must be due to some form of catalytic action. The obvious way to test such an action was to find out whether ornithine would work in lower concentrations just as it did at the relatively high concentrations Krebs had so far employed. Although results of the first experiments conducted with a series of different concentrations were ambiguous, Henseleit was able, by the end of February, to reproduce the ornithine effect at concentrations so low as to show decisively that it must be catalytic.

For a time – it may have been several weeks – Krebs was unable to envision a mechanism for this catalytic action. Guided only by the very general idea that catalysis involves the temporary union between a catalyst and its substrate, he came to believe that the action of ornithine must also be connected to the arginase reaction in the liver, in which arginine is converted to ornithine and urea. But it was difficult to understand the form of the connection, because in this reaction ornithine was one of the products. At some point he realized that ornithine must produce arginine by a different route from the reaction by which arginine produces ornithine. Knowing that the ornithine effect also requires ammonia, Krebs simply constructed a route on paper. To balance the equation he needed, in addition to one molecule of ornithine and two of NH_3, one molecule of CO_2. The result was a cycle composed of two reactions, in which the ornithine transformed to arginine was regenerated by the arginase reaction. Because the first reaction required that four molecules come together at once, he knew immediately that there must be at least one more intermediate step involved. Soon afterward Krebs identified citrulline as the additional intermediate. This expanded cycle he first represented, one year after his initial publication of the discovery, in the form that quickly became familiar:

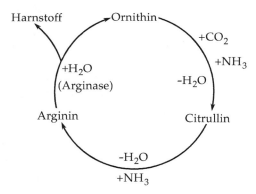

To what extent was the ornithine cycle discovered through logical processes? When reconstructed in greater detail, Krebs's investigative pathway proceeds far less directly to the successful outcome than it appears to in this highly compressed summary. He shifts directions repeatedly; the trend appears sometimes toward one explanatory mode, sometimes toward another. He moves back and forth from efforts to characterize the specific ornithine effect to searches for analogous effects, sometimes digresses to other problems and then returns to urea synthesis. Nevertheless, each experiment in the long series that leads to the discovery can be connected logically to the immediate situation to which his previous experimental pathway has taken him. The logic is not always tight. By his own account, Krebs was prone to try out the effects of many substances without "any too specific ideas." But these trials all took place within a clearly defined set of criteria for identifying an unusual effect to follow up. Moreover, his strategy, trying many different things in the hope that something would "turn up,"[16] was a rational approach to the problem he wished to solve with the means he had available.

In 1988, Deepak Kulkarni and Herbert Simon used my historical reconstruction of Krebs's discovery of the urea cycle as the foundation for a computer program with which they modeled the heuristics Krebs had used during his investigation. The computer was then able to simulate a research pathway, similar to Krebs's own, which recapitulated the discovery.[17] They concluded:

The elucidation of the step-by-step progress of Krebs toward the discovery of the urea cycle shows the discovery being produced by a whole sequence of tentative decisions and their consequent findings, and not by a single "flash of insight," that is, an unmotivated leap. It would appear that whenever we are able to build our models of the discovery process on detailed data, like that provided by Holmes in this instance, scientific discovery becomes a gradual process guided by problem-solving heuristics similar to those used in other intelligent human endeavors.[18]

In a previous commentary on the project of Deepak and Simon I pointed out that they were able to model effectively Krebs's experimental strategy, but did not capture the mental process through which he arrived at the theoretical formulation which integrated his results into the ornithine cycle that constitutes his principal discovery.[19] Was there here, after all, a "flash of insight," an "unmotivated leap," or what Schaffner calls an "illuminationist"[20] experience? According to Krebs's own recollection, there was no single moment when the solution occurred to him. Rather, the situation "became gradually clear."[21] It is, of course, possible that he did have a flash of insight comparable to those described in the famous autobiographical accounts of August Kekulé, Charles Darwin, and a few others, but that this had vanished from his memory, and so from the historical record. In his own published account of the discovery, Krebs presented the event as a logical inference:

> When considering the mechanism of this catalytic action I was guided by the concept that a catalyst must take part in the reaction and form intermediates. The reactions of the intermediate must eventually regenerate ornithine and form urea. Once these postulates had been formulated it became obvious that arginine fulfilled the requirements of the expected intermediate. This meant that a formation of arginine from ornithine had to be postulated by the addition of one molecule of carbon dioxide and two molecules of ammonia, and the elimination of water from the ornithine molecule.[22]

Here the logic of a scientific discovery appears to be fully reconstructed. But we should be cautious about uncritically accepting a reminiscence written forty years later by a scientist who placed a high value on plain, concise expression. If the logic were so transparent, it is difficult to explain how Krebs could have been puzzled for several weeks about the mechanism of the catalytic action. May not the phrase "it became obvious" allude to some long-forgotten psychological event which lay at the heart of the discovery of the ornithine cycle?

Popper wrote that "[t]he question of how it happens that a new idea occurs to a man . . . may be of great interest to empirical psychology; but it is irrelevant to the logical analysis of scientific knowledge."[23] The irony of this claim, when applied to the discovery of the ornithine cycle, is that the psychological situation – exactly how and when, and during what kind of personal activity the solution of his problem occurred to Krebs – cannot be recovered, but the logic of the situation can be fully explicated. The logic involved is not entirely the propositional logic favored by analytical philosophy. The culminating steps in Krebs's solution involved the recognition of a pattern into which all of the elements of the problem "fitted" so compellingly that the evidence for the cycle appeared to him too powerful to doubt.[24] Pattern recognition is, of course, a psychological event, but it does not introduce an "irrational" element into this process of the generation of a discovery. Pattern recognition is as fundamental to the inferences that we draw about the world as any other form of logical structure.

To analyze Krebs's discovery in this way is not to reduce it to a predetermined or routine outcome, lacking creativity or imagination. If the final mental steps through which Krebs reached his solution appear relatively uncomplicated, that is only because the imagination with which he had applied his instrumental, conceptual, and heuristic repertoire to the sustained investigation of the problem of urea synthesis had led him to a position in which the problem had finally become tightly structured. The intellectual gap which he then had to cross was sufficiently narrow that the mental leap he made does not appear mysterious. It is fathomable as a familiar type of experience, one that happens even to ordinary people when they have thoroughly examined a problem placed before them.

The much longer research trail that led Krebs eventually to the discovery of the citric acid cycle began early in 1933, when he took up the broad question, "what are the intermediates in the oxidation of foodstuffs?"[25] Experiments by others on the respiration of isolated tissues dating from the first decade of the century had already established a small list of organic compounds, including a series of dicarboxylic acids (succinic, fumaric, malic, and oxaloacetic acid), as well as the tricarboxylic citric acid, which were capable of increasing the rate of respiration and were therefore potential intermediates. Krebs's initial intention was to test whether these substances would react similarly in his more precise experimental system, and to attempt to identify other compounds that might also increase respiration. Quickly, however, he began to select compounds for study that were incorporated as intermediates in several theoretical reaction schemes under discussion in the field at the time. One of these, known as the Thunberg-Knoop-Wieland scheme, proposed a closed cycle of reactions, in which the four dicarboxylic acids mentioned give rise by decarboxylation to acetic acid, two molecules of which are synthesized to regenerate succinic acid. If this hypothetical pathway could be proven to occur, it would provide a common final pathway for the oxidative metabolism of carbohydrates, fatty acids, and amino acids. Other schemes Krebs tested in this way were a recently proposed ω-oxidation of fatty acids, and the conversion of butyric acid to the "ketone" bodies at the end of the long-accepted β-oxidation series for fatty acids.

Abruptly interrupted in April, when he was dismissed from his post in Freiberg during the Nazi drive to purify the civil service of non-Aryans, Krebs was able by mid-July to resume his investigative trajectory in the biochemistry laboratory of Frederick Gowland Hopkins in Cambridge, England. Continuing to explore the same set of problems, Krebs sometimes went beyond tests of reaction schemes he had found in the contemporary literature. Posing additional possible connections between metabolites included in these schemes, he applied similar quantitative tests to support or rule out his ideas. Several attempts to connect citric acid by means of theoretically plausible reaction chains to the other intermediates led nowhere, because the postulated connecting compounds did not increase the respiration of the tissues. One promising distinction between two possible pathways leading to the formation of acetoacetic acid – one from acetic acid, the other through the β-oxidation of longer-chain fatty acids – faded after further experiments failed to confirm the initial results which had given rise to the idea. In mid-November Krebs broke off his investigation of the whole domain of questions related to the oxidation of foodstuffs that had occupied him for ten months. Feeling that he was "up against a wall,"[26] he returned to the area of nitrogen metabolism from which his earlier successes had come.

His inability to make progress in this investigative field does not mean that Krebs had pursued these problems less resourcefully than he had pursued the brilliantly successful investigation of urea synthesis during the previous year. His style remained the same. Although he shifted from one subproblem to another frequently, sometimes testing coherent reaction schemes, sometimes searching more diffusely for clues that would give direction to his endeavor, each experiment he tried can be connected logically to his previous experiences or to the views propounded in the current literature of the field. Even his decision to give up on these problems seems a rational conclusion that his chances for solving them were less than his chances for making further advances in his other major subfield of activity.

During the next two years Krebs did make considerable progress in the area of nitrogen metabolism, discovering, among other things, the synthesis of glutamine and part of the pathway of uric acid synthesis in birds. In the spring of 1936 he reexamined a scheme for amino acid synthesis proposed twenty-six years earlier by Franz Knoop, which involved an anaerobic condensation reaction between a pyruvic acid, another ketonic acid, and ammonia. During the course of his experiments Krebs found that ammonia was not necessary to the reaction, a result which led him to examine anaerobic reactions between two ketonic acids in which an oxidation is balanced by a reduction. Called "dismutations," reactions of this type were prominent in the recently formulated Embden-Meyerhof pathway of glycolysis, or anaerobic carbohydrate metabolism, and were widely discussed in biochemistry at the time. During the next few months Krebs produced evidence, from the stoichiometric ratios of substances consumed and produced, for several reactions of this type taking place in liver and other tissue slices. Believing that the reactions might connect the anaerobic and aerobic stages of carbohydrate metabolism, he pursued this investigative pathway energetically.

This stage in Krebs's experimental pathway illustrates how subtly interconnected the contexts of discovery and of justification often are, and how a sequence, each step of which is connected logically to the preceding steps, can lead the investigator in unexpected directions. What began as an experimental test of a theory of amino acid synthesis metamorphosed into the generation of a new set of hypotheses about hitherto unknown stages in the oxidation of carbohydrates. These steps formed an intellectual bridge across which Krebs moved from the field of nitrogen metabolism back to the broad questions about the oxidation of foodstuffs that he had abandoned more than two years before. Thus Krebs reasoned his way, step by step, along a track whose direction often shifted in ways that reason could not foretell.

By early July, Krebs believed that he had acquired sufficient evidence to put forward, in an article in *Nature*, a scheme in which a sequence of three reactions connected the anaerobic dismutation of pyruvic acid, through similar dismutations involving other ketonic acids, to the formation of succinic acid. Because succinic acid had long been identified as central to respiratory oxidation, and connected to the dicarboxylic acid sequence in the Thunberg-Knoop-

Wieland scheme and other variants, Krebs was, in effect, proposing links that would connect the separate existing metabolic sequences into complete pathways of carbohydrate and fatty acid breakdown.

Through his daily laboratory record, one can follow almost all of the reasoning and experimental exploration through which Krebs built up the picture he presented here. The occasional gaps left in the trail of his reasoning do not suggest that any "illogical" mental leaps occurred, but only that he did not always leave sufficient traces on paper to enable every step in his thinking to be reconstructed. We see him reasoning his way to a hypothesis which was not yet a discovery, because he acknowledged that two of the three reactions in question were still only "tentatively formulated." As he continued his investigation these formulations became more rather than less tentative.

Through August and September Krebs widened his search for dismutation reactions that might connect the intermediates of the anaerobic and aerobic phases of carbohydrate metabolism. He extended his scope also from mammalian tissues to bacterial metabolism. In addition to a lengthening list of reactions between each of the keto-acids that had been implicated in metabolic reactions, he began testing the dicarboxylic fumaric acid, alone and in various combinations. In doing so he was influenced by the recent discovery of Albert Szent-Gyorgyi and his collaborators in Hungary, who had found that fumarate added in catalytic quantities to minced pigeon breast muscle tissue restored its respiration. Szent-Gyorgyi had propounded, on the basis of this and related experiments, that succinic, fumaric, and oxaloacetic acid constituted a "carrier system" that transferred hydrogen removed from substrate molecules to the cytochrome transport system. In the process, succinic acid was oxidized to fumaric, then to oxaloacetic acid, which was then reduced directly to succinic acid, completing a cycle. Krebs considered Szent-Gyorgyi's general concept of a cyclic carrier system to be significant, but its particular feature of an "overreduction" of oxaloacetic to succinic acid seemed wrong to him, and he spent part of his time looking for alternative ways to understand that part of Szent-Gyorgyi's scheme. During this time, therefore, Krebs seemed to view his own ongoing experimental object from two alternative perspectives – as a sequence of dismutation reactions constituting a metabolic pathway (an extension of the Embden-Meyerhof pathway), and as a set of carrier systems and reactions in the mode of Szent-Gyorgyi's fumarate system. He seemed also to oscillate between the search for a single pathway among the alternate reaction possibilities he was exploring, and the idea that there may actually be a network of analogous parallel or intersecting pathways involved. Thus the reasoning behind the experiments that he pursued with special intensity during these months was open-ended, somewhat loose, but entirely rational in the context of his experimental situation. His method allowed multiple hypothetical possibilities, among which he could not, for the moment, choose. All were, however, based on the same fundamental conceptual assumptions, and all were tested by the same experimental criteria. He could only assume that by pursuing these investigative leads with persistence, he would eventually encounter experimental results that would rule out all but the correct alternatives. Thus here, too, the generation and testing of theories were inextricably embedded in one another. The same experiments that generated or eliminated candidates for the sequence or family of pathways he sought were expected to hone and sharpen a somewhat diffuse theoretical structure until he could arrive at a tightly connected reaction scheme.

Among the dismutation schemes that Krebs had Johnson test were two – malic + pyruvic acid, and oxaloacetic acid + pyruvic acid – that gave rise to citric acid. Repeatedly since 1933, Krebs had made transient efforts to connect citric acid to known reaction sequences, but with little success. Now, for the first time, he found a more promising lead. The logic was a direct extension of the dismutation scheme with which he had been seeking all summer to connect the products of glycolysis to succinic acid and the oxidative breakdown of carbohydrates. At almost the same time, Carl Martius and Franz Knoop published chemical evidence that oxaloacetic and pyruvic acid yield citric acid. Citric acid did not lead, however, back to any of the other intermediates, so that to find a pathway from them to citric acid only complicated Krebs's search for a larger coherence among the family of pathways he was exploring.

During October and November, Krebs gave several lectures in which he presented the current state of his thinking about the "oxidative breakdown of carbohydrates." Less confident than in July that he had identified the primary reactions giving rise to succinic acid, he now offered several alternative reactions that might connect glycolysis to the formation of that substance. Moreover, he admitted now that not all of the pyruvic acid oxidized went through succinic acid. In bacteria, a second path was available through the direct formation of acetic acid. In animal tissues, the second pathway was probably the synthesis of citric acid. Krebs emphasized, however, that the formation of citric acid through the breakdown of pyruvic and malic acid, "appears to be a side reaction only," and that "we know nothing about the products of oxidation of citric acid."[27]

Krebs continued during the winter and early spring of 1937 to gather more evidence for the various dismutation reactions that he had by now identified. Multiplying his controls, he endeavored to gather more rigorous quantitative data concerning both the rates of the reactions and their balance sheets. As he did do, his overall scheme tended further to dissolve into a group of partial schemes. He no longer claimed to have found the specific reaction sequence connecting glycolysis with the aerobic breakdown of carbohydrates, but only a group of analogous dismutation reactions which "appear to play a role in the course of the normal oxidative breakdown of carbohydrates." Moreover, the particular reactions that seemed prominent in some organisms, such as bacteria, were insignificant or missing in others, such as animal tissues. The goal of linking together all the known pieces of the picture that had seemed so close during the preceding summer now appeared to elude his grasp.

Sometime during the third week in April, Krebs came across a paper by Martius and Knoop proposing a reaction mechanism for the decomposition of citric acid. Derived by Martius through his understanding of organic reaction mechanisms and an analogy to the physiological reactions of the dicarboxylic acids, and tested qualitatively by means of enzymatic tissue extracts, the proposed sequence was citric acid → cis-aconitic acid → isocitric acid → oxalosuccinic acid → α-ketoglutaric acid. The last acid in the sequence, Martius pointed out further, was known to give rise to succinic acid by decarboxylation, thus connecting the sequence to the long-known sequence of the dicarboxylic acids. When he read this paper, Krebs must have seen almost at once that it solved for him the mystery of what happened to the citric acid whose metabolic formation he had included in the family of dismutation reactions he had proposed during the preceding fall. Its synthesis was no side reaction, because the new decomposition pathway reconnected citric acid to what he and others had assumed to be the main pathway of carbohydrate oxidation, through succinic acid. Quickly he had Johnson carry out an experiment to test the effects of citric acid on the respiration of minced heart muscle tissue, adding malonate to inhibit the further oxidation of the succinate that might be formed. A large increase in the quantity of succinate dramatically confirmed that Martius's mechanism occurred in the tissue at a rate sufficient to suggest that it could lie on the main pathway of respiratory oxidation. By this time the outlines of a cyclic sequence of reactions must have been clear to Krebs. He thereafter directed his efforts and those of Johnson to gather further confirming evidence. It required less than three weeks to perform the further experiments necessary for the evidence to become convincing enough for Krebs to write a preliminary paper proposing the "citric acid cycle" as the main "pathway through which carbohydrate may be oxidized in animal tissues."[28]

As in the case of the ornithine cycle, Krebs recalled no specific moment at which the idea of the citric acid cycle first occurred to him. It was possible, however, from the logic of the situation, to reconstruct the circumstances under which it occurred to him, and from the record of his experiments and the date of the publication of the paper of Martius and Knoop to locate the approximate time the insight took place. Once that is done, the "logic" of the discovery is nearly transparent. Once again, the only aspect of the discovery that eludes us is the precise "psychological" dimension of the experience.

Out of the hundreds of experiments that Krebs and his assistants performed along the trail that led, between 1933 and 1937, to the discovery of the citric acid cycle, there are only a handful for which Krebs's reasoning and the logic connecting them to his investigative pathway up to that point cannot be reconstructed (or for which a possible reconstruction has eluded me); and in these cases the problem seems only to be that some crucial bits of information are missing, not that some unfathomable mental leap had taken place. My motive in constructing an "unbroken" investigative pathway through these years was the conviction that the springs of creativity in experimental science lie in the daily interplay between thought and operations, and that if nothing were left out, the course that a scientist followed to a major discovery would no longer appear to be inscrutable or to contain "irrational" elements. It may be left to philosophers to inquire whether there is a "deep logic" underlying Krebs's investigative habits, but I believe I can claim that his pathway has been rendered "intelligible to reason."

One explanation for the fact that Krebs's solutions to the problems of urea synthesis and of the pathway of oxidative carbohydrate metabolism can be made fully intelligible to reason is that the solutions conformed to strict existing criteria for the formulation of a sequence of metabolic reactions: that is, in addition to the immediately relevant experimental data, they incorporated known structural formulas for the organic compounds involved, and long-standing rules for balancing chemical equations. The fact that Krebs's solutions broke no such rules may tempt some to regard them not as true discoveries, but merely as solutions to the puzzles arising in a normal science.

"Discovery commences," Thomas Kuhn asserted,

with the awareness of anomaly, i.e., with the recognition that nature has somehow violated the paradigm-induced expectations that govern normal science. It then continues with a more or less extended exploration of the area of anomaly. And it closes only when the paradigm theory has been adjusted so that the anomalous has become the expected. Assimilating a new sort of fact demands a more than additive adjustment of theory, and until that adjustment is completed – until the scientist has learned to see nature in a different way – the new fact is not quite a scientific fact at all.[29]

In Kuhn's treatment, the "emergence of discoveries" through anomalies shades into the more radical emergence of new theories when the prevailing paradigm cannot be adjusted to resolve the anomaly. The advent of discoveries shares, therefore, something of the qualities that Kuhn ascribed to the arrival of a new paradigm:

More often . . . the new paradigm, or a sufficient hint to permit later articulation, emerges all at once, sometimes in the middle of the night, in the mind of a man deeply immersed in crisis. What the nature of that final stage is . . . must here remain inscrutable and may be permanently so.[30]

Normal science, Kuhn stressed, "does not aim at novelties of fact or theory and, when successful, finds none."[31]

Despite his acknowledgement that solving the puzzles of normal science often requires exceptional skill and ingenuity, that it attracts passionate commitment, and that it challenges the minds of the greater scientists; and despite the fact that he occasionally (and inconspicuously) applied the word *discovery* also to finding how the "rules" of an existing paradigm "could successfully be applied" to solve a puzzle of normal science,[32] the strong impression that Kuhn's *Structure* has left is the contrast between the novelty of revolutionary science, and its absence in normal science. His characterization of normal science as "mop-up work"[33] has most often been echoed when other authors summarize his views.

The predilection for counting as discoveries only those events that somehow break the rules of the science within which they emerge is evident also in the views of other philosophers – for example, in Carl Hempel's statement that a logic of discovery would have to "provide a mechanical routine for constructing, on the basis of given data, a hypothesis or theory stated in terms of some quite novel concepts which are nowhere used in the description of the data themselves."[34]

Whether they consciously adopt, or resist, Kuhn's account of the development of science, historians and philosophers alike have tended since the publication of *Structure* to absorb his dichotomy between normal and revolutionary science. That dichotomy makes it more difficult to understand the logic of discovery by appearing to remove from that category most of the events that scientists themselves count as discoveries. Despite the fact that it did not break any of the rules that had previously guided the search for metabolic reactions, the Krebs cycle is described in Alfred Lehninger's authoritative textbook of biochemistry as "the most important single discovery in the history of metabolic biochemistry."[35] Such recognition of its stature is sufficient warrant for us to regard the investigative pathway which led to the Krebs cycle as an exemplar which can help us to elucidate the logic of discovery.

Even without the Kuhnian influence, discovery has often been associated with unexpected or surprising observations. Hanson's effort to devise a logic of discovery, for example, begins with the encounter with "some surprising, astonishing phenomena" which would not be surprising if a certain type of hypothesis could be brought to bear on it.[36] If we accept the ornithine and citric acid cycles as true discoveries, however, we see that the element of surprise may or may not play a conspicuous role. The ornithine cycle was the outcome of Krebs's response to the "ornithine effect," an entirely unexpected difference between the effect of this particular amino acid and the other amino acids he had tested, on the rate of urea formation. Along the much longer trail that led Krebs to the citric acid cycle, there were no comparable surprising observations that determined the subsequent course of the investigation. There was instead a prolonged process of trying multiple possible solutions within the rules that defined the problem; of reasoning by analogy to a recently discovered pathway; of schemes built out of early experimental explorations that partially dissolved in the solvent of further experiments, but left elements that were incorporated into subsequent versions; of a key piece of the solution worked out elsewhere appearing in a timely publication; until finally a pattern emerged that fitted so well the tests that could be devised that Krebs knew he had found the answer he had so long sought. If we were to study other investigative pathways as fully documented as these, and if we made the effort to reconstruct them at an intimate level of detail through their entire trajectory, we would surely find similar patterns leading to the discoveries, large and small, that make up the history of experimental biology.

NOTES

1. Bruno Latour and Steve Woolgar, *Laboratory Life: The Construction of Scientific Facts,* second ed. (Princeton: Princeton University Press, 1986).
2. Steven Shapin and Simon Schaffer, *Leviathan and the Air Pump: Hobbes, Boyle, and the Experimental Life* (Princeton: Princeton University Press, 1985).
3. Ian Hacking, *Representing and Intervening: Introductory Topics in the Philosophy of Natural Science* (Cambridge: Cambridge University Press, 1983), pp. 262–275.
4. Hans-Jörg Rheinberger, *Toward a History of Epistemic Things: Synthesizing Proteins in the Test Tube* (Stanford: Stanford University Press, 1997).
5. Robert P. Crease, *The Play of Nature: Experimentation as Performance* (Bloomington: Indiana University Press, 1993).
6. Pat Langley, Herbert A. Simon, Gary L. Bradshaw, and Jan M. Zytlow, *Scientific Discovery: Computational Explorations of the Creative Process* (Cambridge, MA: MIT Press, 1987).

7. Kenneth F. Schaffner, *Discovery and Explanation in Biology and Medicine* (Chicago: University of Chicago Press, 1993).

8. Karl R. Popper, *The Logic of Scientific Discovery* (New York: Science Editions, 1959), pp. 31–32.

9. Thomas Nickles, "Introductory Essay: Scientific Discovery and the Future of Philosophy of Science," in *Scientific Discovery, Logic, and Rationality*, ed. Thomas Nickles (Boston Studies in the Philosophy of Science, vol. 56) (Dordrecht: D. Reidel, 1978), pp. 1–60.

10. Schaffner, *Discovery and Explanation*, pp. 97–99.

11. Popper, *Logic*, pp. 29, 40.

12. Schaffner, *Discovery and Explanation*, pp. 8–63.

13. Ibid., p. 43.

14. Ibid., p. 97.

15. Frederic Lawrence Holmes, *Hans Krebs: Volume 1, The Formation of a Scientific Life, 1900–1933* (New York: Oxford University Press, 1991); Frederic Lawrence Holmes, *Hans Krebs:* Volume 2, *Architect of Intermediary Metabolism, 1933–1937* (New York: Oxford University Press, 1993).

16. Hans Krebs and F. L. Holmes, personal conversations, August 1 and 4, 1978.

17. Deepak Kulkarni and Herbert A. Simon, "The Processes of Scientific Discovery: The Strategy of Experimentation," *Cognitive Science* 12 (1988): 139–175.

18. Ibid., p. 174.

19. Frederic L. Holmes, "Research Trails and the Creative Spirit: Can Historical Case Histories Integrate the Short and Long Timescales of Creative Activity?" *Creativity Research Journal* 9 (1996): 248.

20. Schaffner, *Discovery and Explanation*, p. 19.

21. Hans Krebs and F. L. Holmes, personal conversation, August 1, 1978.

22. Hans Krebs with Anne Martin, *Reminiscences and Reflections* (Oxford: Oxford University Press, 1981), pp. 56–57.

23. Popper, *Logic*, p. 31.

24. Hans Krebs and F. L. Holmes, personal conversations, September 7, 1976, August 1, 1978.

25. Ibid., May 2, 1977.

26. Ibid., May 4, 1977.

27. Hans Krebs, "The Oxidative Breakdown of Carbohydrates," typewritten manuscript, H.100, Krebs Collection, Sheffield University Archives.

28. H. A. Krebs and W. A. Johnson, "The Role of Citric Acid in Intermediate Metabolism in Animal Tissues," typewritten manuscript dated June 10, 1937, Krebs Collection, Sheffield University Archives. This paper, submitted to *Nature*, was not accepted for immediate publication. Two weeks later Krebs submitted a longer manuscript, which included a few additional experiments performed in the interval, to the journal *Enzymologia*.

29. Thomas S. Kuhn, *The Structure of Scientific Revolutions,* second ed. (Chicago: University of Chicago Press, 1870), pp. 52–53.
30. Ibid., pp. 89–90.
31. Ibid., p. 52.
32. Ibid., p. 39.
33. Ibid., p. 24.
34. Quoted in Schaffner, *Discovery and Explanation,* pp. 48–49.
35. Alfred Lehninger, *Principles of Biochemistry* (New York: Worth Publishing, 1982), p. 443.
36. Quoted in Schaffner, *Discovery and Explanation,* pp. 12–13.

Chapter 9

What Do Population Geneticists Know and How Do They Know It?

R. C. LEWONTIN

In the movement of philosophers of science away from a preoccupation with physics and with the immense growth of attention paid to problems in biology, there has been what might seem a disproportionate interest in evolution and evolutionary genetics. But this concentration on one field of biology is a reflection of the uniquely difficult epistemological issues raised. Much of the earlier philosophical fascination with physics, and especially with relativity theory, particle physics, and quantum mechanics, was a consequence of the apparent contradiction between the structure of physical theory and what seemed a rational and commonsense understanding of the world of gross phenomena, a contradiction that was perceived by physicists themselves, for whom epistemological questions were always at the forefront. In like manner, the current emphasis placed by philosophers on epistemological and methodological issues in evolution and evolutionary genetics is a reflection of the self-conscious grappling with these problems on the part of biologists. Before professional philosophers began concerning themselves with the problems of the field, evolutionary and population geneticists had been preoccupied by methodological and epistemological difficulties in a quite explicit way, although not always with the greatest precision of thought. Indeed, there is some imprecision in the definition of "philosopher" as opposed to "biologist" in these matters.[1] (See Lewontin 1974 for a review of the situation just prior to the flowering of interest

This essay has benefitted immensely from a very critical reading given to an earlier version by Lisa Lloyd, to whom I am immensely grateful. I am also very grateful to the editors of this volume for their light but unerring editorial hand.

on the part of professional philosophers.) The source of the philosophical interest in evolutionary genetics has not been, as it has been in physics, the fundamentally paradoxical ontological properties of the objects of inquiry. Organisms seem to behave in quite commonsensical ways, and their history does not pose problems that seem to confound rationality. The philosophical issues arise not out of a lack of intuitively reasonable explanations of what we observe but, on the contrary, from a superfluity of such explanations. The problem is how to decide in any particular case which of the reasonable explanations to believe. It is the epistemologist's paradise.

THE FORM OF EVOLUTIONARY EXPLANATIONS

To understand how such a superfluity arises, we need to remind ourselves of the structure of evolutionary theory in its modern genetical form. The shape of explanation in population genetics arises from two features of our understanding of the evolutionary process itself. On the one hand, the process is a consequence of a large set of universal basic biological mechanisms that are not in themselves contingent in the way in which they enter into causal chains but only in their quantitative force in each instance. Some of this quantitative contingency arises from properties of the organisms themselves – for example, whether they reproduce sexually – but there are external forces as well, coming from autonomous changes in physical nature or in other species. On the other hand, as a result of the quantitative contingency of the forces, of the stochastic nature of their actual operation, and of the particular nature of the organisms on which the mechanisms work in each instance, evolution is an irreducibly historical process. That is, the current state of living organisms is a consequence of contingent events in the past, and the future state of the living world is constrained in its possibilities by the current collection of actually existing organisms. Moreover, the collection of organisms is not at some global equilibrium or steady state; present life bears the marks of the particular historical trajectory through which it has passed. Thus the explanation of the present state of organisms and the prediction of their future state, including the deliberately manipulated evolution of species by controlled breeding, must be considered in terms of universal mechanisms whose operation is historically contingent, both because of the importance of initial conditions and because of autonomous externalities that have their own history.

The Darwinian variational theory of evolution begins with the fact that individual organisms within a species vary in their properties and that such variation is to some degree heritable, so that offspring resemble their parents more than they resemble unrelated individuals. Evolution within a species consists in a change in the ensemble of individuals that comprise that species, through an enrichment in the proportion of some variant types and a reduction in the proportion of others. This relative enrichment is a consequence of the fact that some individuals leave more offspring than others and that their properties are heritable. Differences between species in space and time are the result of the conversion of differences between individuals within species into differences between groups as a result of the different ensemble changes in different populations at different times. This skeletal explanation seems unproblematic, since it is evident that variation among organisms is, at least in part, heritable, and that every individual in a population does not leave the same number of offspring. The epistemological problems arise from the details of mechanism. In particular, we need to consider the variations in quantitative effect of the elementary forces, the nature of the coupling between the forces, and most important, the sensitivity of those forces to particular internal and external circumstances.

Evolutionary genetics has a peculiar epistemological "texture" that cannot be analogized to physics or even to other branches of biology. These disciplines are bad models for understanding the philosophical problems of the field. So, it is *not* within the problematic of population geneticists to discover the basic biological phenomena that govern evolutionary change, as it was for nuclear physics to discover universal forces between nuclear particles. The basic phenomena are already provided to population genetics by biological discoveries in classical and molecular genetics, cell biology, developmental biology, and ecology. Nor is it within the problematic of *observational* population genetics to discover the ways in which the operation of these causal phenomena can interact to produce effects. The elucidation of the structure of the network of causal pathways, and of the relation between the magnitudes of these elementary forces and their effects on evolution, is an entirely analytic problem. But neither is population genetics like engineering or applied physics. It cannot be the task of population genetics to fill in the particular quantitative values in the basic structure that will provide a correct and testable detailed explanation of what has happened in any arbitrary case. As will be argued below, we lack the necessary observational power and, in practice, will always lack it. Rather, the task of population genetics is to make existential claims about both outcomes of evolutionary processes and about significant forces that contribute to those outcomes. As it has turned out, both for epistemological and ontological reasons, there is an inverse relation between the degree of specificity of these existential claims and the size of the domain to which they apply. So, for example, population geneticists say that *virtually all* populations show a lot of enzyme polymorphism, that natural selection is important in determining the pattern of these polymorphisms in *many* cases, but that selection in favor of heterozygotes is *very rarely* the cause of the polymorphism.

Like other scientists, population geneticists have been educated to believe in a naive univocal model of science, a model that takes the discovery of universals to be the final validation of scientific inquiry and an accurate quantitative evaluation of all relevant causal variables to be the mark of a truly scientific explanation. They are then dissatisfied with a science that says that "*x* can happen" or that "*y* sometimes happens" or even that "*z* often happens." Moreover, in the explanation of a specific case they regard it as a defeat to be able to say only that "*x* could not have been the cause of the observations, *y* could have been a significant factor, and *z* certainly played an important causal role." Yet for evolutionary phenomena, with so many weakly determining and interacting causal pathways and with a dependence on historical contingency, that is the best that can be done.

THE FORCES OF GENETIC VARIATION AND CHANGE

Of the elementary forces, there are, first, the processes that give rise to heritable variation. These include entirely internal mechanisms like mutations from nucleotide substitutions, deletions, duplications, and insertions, whose rates are largely determined by internal cellular conditions and are relatively insensitive to environmental circumstances. But there is also recombination, whose rate is partly internally determined by the chromosomal distribution of genes, but whose population effect depends strongly upon how much mating there is between close relatives, how much selfing has occurred in species with mixed selfing-outcrossing systems, and, in organisms with both sexual and asexual reproduction, how much of each kind of reproduction has taken place. These factors vary considerably with environment and are strongly dependent on the actual natural circumstance. In addition there is the introduction of variation within local populations by immigrants, a major source of potential variation that is extremely sensitive to temporal and geographic circumstances.

Second, among the forces there are several processes that result in changes in the frequencies of particular genetic variations within populations, of which natural selection is only one. Asymmetrical inheritance mechanisms like meiotic drive are very powerful in changing the frequencies of whole linked blocks of genes, provided the genetic variant responsible for the drive is still segregating within populations. But a driven chromosome that has completely replaced its homologue in a population is no longer a drive element, since there is no alternative to be driven. This is simply an example of the general self-negating nature of driving forces in genetic evolution. Selection (or any other directional driving force that leads to the replacement of one variant by another) can exist only when there is an alternative to be selected. Natural selection does not occur at loci that have no genetic variation in the species. Moreover, we must distinguish between selection *for* a gene and selection *of* a gene. Because of genetic linkage on chromosomes, if an allele at some locus is swept to homozygosity by natural selection operating on the properties of that locus, any alleles at loci closely linked to it that happen by chance to be in linkage disequilibrium with the selected locus will also be driven to high frequency or homozygosity by genetic "hitchhiking". This will occur even if these accidentally linked alleles are at a small intrinsic selective disadvantage or of no selective significance at all.

There is an immense variety of scenarios of direct natural selection. These include positive and negative purifying selection, balanc-

ing selection that favors intermediate allele frequencies, and fre-
quency- and density-dependent selection. All of these are usually
dependent both for their direction and intensity on particular environmental circumstances and on the current state of the genotype
at other loci. It is these last dependencies that make initial conditions
so powerful in influencing the course of evolutionary change.
Whether a new mutation will be favorable, unfavorable, or without
selective importance depends on the environment of the species into
which it is introduced; but the environment of a species is itself a
consequence of the biology of the species, which is a manifestation of
its entire genotype. The fitness effect of a given mutation in the gene
for alcohol dehydrogenase depends upon the role that alcohol plays
in the metabolism of the organism, which in turn is different for an
insect living on fermenting fruit than for one living, say, by sucking
blood. Putting aside environment, genes and gene products interact
in development by DNA-protein and protein-protein binding, so
that the cellular effect of a mutation at one locus is dependent on the
genotype at other loci.

A long-term increase or decrease in frequency is not uniquely the
result of the operation of a directional force. In any real finite population of organisms different individuals leave different numbers of
offspring, and the very finiteness of population size, added to the
stochastic nature of genetic segregation in heterozygotes, produces
changes in allele frequency for every gene. Such random drift fluctuations result eventually in the fixation of alleles at a rate that depends on the breeding structure and demography of the population.
But these, in turn, vary from time to time and place to place as a
consequence of fluctuations in specific environmental conditions.

Beyond the externally imposed contingency of the magnitude of
the various forces operating on population variation, there is an
internal coupling among the forces. Because populations are finite in
size and because genetic segregation in heterozygotes is stochastic,
even "deterministic" forces such as selection favoring new advantageous alleles operate in real populations in a noisy manner. The
noise is a consequence of the random fluctuation of allele frequencies
from generation to generation in a population of finite size. If the
noise level is great enough, the effect of the signal is lost. What
matters is the signal-to-noise ratio, and since the random drift noise

is proportional to the reciprocal of population size, N, then it is the product Nm or Ns or $N\mu$ that determines whether a particular intensity of migration, m, or selection, s, or mutation, μ, is actually effective in the evolution of the population. Generally if these products are smaller than unity, the effect of the deterministic force is not felt and they remain undetected by the population, which evolves as if only genetic drift were operating. But if the products are larger than unity there may be long-term effects on population composition even when the magnitudes of the directional forces are very small. So, in a population of size 10^6, a selective disadvantage of even 1 in 10,000 will be effective in preventing a mutation from rising to any significant frequency. The converse is not true, however, and a single newly arisen mutation with a selective advantage of 1 in 10,000 has a probability of only about 1 in 5,000 of successfully invading the population, irrespective of population size; so such a mutation must occur over and over if it is eventually to characterize the species. Another important consequence of the "product rule" is that the exchange of even a single migrant among populations each generation is enough to prevent them from differentiating by random drift alone. Because m, the migration rate, is the number of migrants as a proportion of the total population size, N, then the product, Nm, is the absolute number of migrant individuals.

There are, moreover, secondary effects of the coupling between mechanisms. For example, a weak but pervasive cause of homozygosity at unselected loci is the constant low level of purifying selection, so-called "background selection." In rejecting newly arisen deleterious mutations this selection reduces the effective breeding size of the population and thus increases the rate of genetic drift.

Finally, account must be taken of the fact that the rates of temporal change that are consequent on the operation of the various forces are, for the most part, quite small. The change in allele frequency in a single generation is, in general, roughly equal to the magnitude of the force multiplied by a function (always less than unity) of the frequency of the allele in question. So, for example, the expected increase in one generation of a selectively favorable allele whose frequency is q and whose selective advantage is s is given by:

$$\Delta q = sq(1 - q)$$

Thus, an allele whose selective advantage is say, $1/1{,}000$, cannot increase by more than .00025 per generation when it is intermediate in frequency, and much more slowly when it is either rare or very common (q close to 0 or 1). The rate of change from repeated mutation is always less than the mutation rate itself, which is generally of a smaller order than 10^{-5}, and the rate of fixation of alleles by genetic drift is proportional to $(1/N)$, the reciprocal of population size. Thus, except under unusual circumstances – the influence of an extremely strong selective agent such as an insecticide, for example, or a catastrophic disease to which a few individuals are genetically resistant; or the founding of a new population from a very small number of migrants – the temporal rates of change of genetic composition are likely to be extremely small.

THE VARIED PROBLEMATICS OF EVOLUTIONARY GENETICS

The epistemological difficulties that arise out of the actual nature of evolutionary processes depend on what we conceive the problematic of evolutionary genetics to be. At various times population geneticists have claimed different programs of investigation, while not always being conscious that these represent different problematics with different methodological and epistemological problems, or that their own practice may be constructed around a different problematic than they imagine. These programs vary in the degree of their ambition, from a maximal inferential program meant to give a correct biological explanation of any and all observed evolutionary differentiation, down to a minimal deductive program that provides the rules for recognizing acceptable and unacceptable explanations, without reference to any particular observed case. In the end, this minimal program may be the only one that is definitively satisfied.

The maximal inferential program of population genetics is to give a correct account of the forces of mutation, migration, selection, and breeding structure that have led to any and all observed patterns of genetic variation within populations, between populations of the same species, and even between species that are sufficiently closely related that their divergence is a simple extension of intraspecific population differentiation of their common ancestral form. Moreover, this account, at least for selection, must be framed in biological rather than purely numerical terms, using as explicans the physiological, behavioral, or anatomical properties of the genotypes. While population and ecological geneticists sometimes speak as if this were *the* program of explanation, it is obvious that it is impossible of realization because it would involve the ability to know all the relevant details of the biology and natural history of any arbitrarily chosen species and of any arbitrarily chosen genetic variation within that species. But the natural history of most species makes them inaccessible to the kind of in situ investigation that would be needed, and the physiological action of the products of most genes either is incompletely known, or the phenotypic variation associated with the genetic variation is so slight as to be beyond practical limits of observation. Population genetics cannot seriously accept the challenge of the maximal program when the rest of biology remains in an incomplete state. The role that the maximal program plays in the field is to provide an idealized pattern of investigation for the actual realization of research. Yet evolutionary biologists are not always prepared to admit that they cannot carry out the program, feeling, perhaps, that to admit that most cases are beyond their reach compromises evolutionary biology, and even biology as a whole, as compared with the model presented by the physical sciences. But of course no physicist pretends that a quantum mechanical description of an apple is possible, or even desirable, and aeronautical designers still have to use real or virtual wind tunnels to know whether their designs will get off the ground. The result of this unrealistic view of what can be known, in evolutionary biology broadly, has been the proliferation of plausible but uncheckable stories of the spread of heritable characters in the past leading to species as we now see them.

At the other extreme is the minimal deductive program that comprises theoretical population genetics. Its purpose is to provide a rigorous network of relationships between the causal forces and their outcomes at the genetic level. The form of this program is to produce purely analytic results of an "if, then" form that can be used to demarcate the allowable from the unallowable claims of explanation. "If there are many populations of size N, exchanging migrants at rate m, and there is no selection, then the steady state variance among populations in the frequency of an allele with mutation rate μ will be . . ." As an analytic program this has no truly epistemological problems, only questions of methodological ingenuity. Many mathematical systems have no explicit solution in closed form, so there are issues of the accuracy of numerical approximations or of the adequate numerical exploration of a parameter space of high dimensionality; but these are problems of art (see Lewontin, Ginsburg, and Tuljapurkar 1983 for an example).

The conflict that arises from a commitment in principle to an unrealizable maximal inferential program has given rise to a set of research enterprises that represent successive reductions in the ambition of the maximal program. In one direction, this reduction attempts to maintain the detail of explanation but sacrifices the generality of the domain over which the explanation applies by choosing a particular variation in a particular species as a model case. The other direction is to accept a lower and lower level of specification of the questions to be asked, merging more and more of the details of explanation into large classes of distinguishable accounts that are meant to have some generality over organisms and traits. So one may sacrifice the demand for biological mediation and settle for an estimate of the numerical values of the reproductive fitnesses of genotypes in nature, although even this degree of specification is very difficult to satisfy except in special model species. A further reduction in detail is accomplished by settling for the order relations of parameters (heterozygotes more fit than homozygotes, or population I much larger than population II, for example) without providing numerical estimates. A yet further coalescing of alternatives drops any attempt to characterize magnitudes and asserts simply that selection, for example, has or has not operated to influence the genetic structure of the population. Indeed, much of present-day molecular population genetics is at this lowest level of problematic.

The research enterprises involving model cases are paradoxical. A model natural heritable variation is chosen whose genetic basis is simple and whose phenotypic manifestations are likely a priori to have large effects on the organism. This genetic variation must occur in a model species whose natural history is such as to make observations of the biology of the system in nature feasible, while at the same time allowing a variety of deliberate perturbations, such as breeding tests or transplantation of individuals from one locality to another. The locus classicus of this model system is the study of the variation in shell color and banding in natural populations of the snail *Cepaea nemoralis* by Lamotte (1951), in which the forces of breeding structure, migration, mutation, and selection were explicitly estimated in nature, and various hypotheses of the biological mediation of selective differences were considered. Snails could be bred in the laboratory to determine the simple genetic basis of the traits, and it was possible to mark snails in nature and to move them around. A somewhat less ambitious form of the model program also relaxes the demand for a specific biological mediation (although a plausible biological story is always welcome). All that is required is some quantitative estimate of the strength of the various forces operating on the model system, without trying to cash these out in terms of natural history. The most complete program of this sort was the study by Dobzhansky and his colleagues of the chromosomal inversion systems in natural populations of *Drosophila*. The forty-three papers in the Genetics of Natural Population series produced by this group contain no attempt to provide a biological explanation of the apparent strong natural selection operating to maintain polymorphism, to cause regular seasonal cycles in inversion frequencies, and to cause geographical differentiation.

What is paradoxical in these programs, which have made up a significant fraction of the efforts of experimental evolutionary geneticists, is that it is unclear what larger problematic is served. Suppose it were possible to give an explanation of a case that was completely satisfactory to all concerned (neither Lamotte nor any later investigator of biological model systems has succeeded in giving an uncontroversial explanation of a particular case). What hypothesis about evolutionary dynamics would then have been tested? If all the available explanations were rejected, that would not cause us to reject the structure of evolutionary genetic theory, but would only suggest that some biological fact had been left out of account.

It is sometimes said that such model cases are meant to be illustrative – but illustrative of what? Of natural process or of method? How much is gained in our understanding of evolutionary processes by showing that *Cepaea* shell patterns or *Drosophila pseudoobscura* chromosomal inversions are evolving under some particular combination of selection, migration, and genetic drift? A great deal would be gained if these model cases could be argued to be somehow typical of a large class of natural situations. But, unfortunately, they have been chosen precisely because the genetic variations have major effects on phenotype (as in the shell patterns) or comprise a large fraction of the genome (as in the case of the inversions), so they are likely to be unrepresentative of the usual genetic variation within species and divergence between species. On the other hand, such studies may be meant explicitly as illustrative of the power of a method, as in the extremely elegant estimates of mating patterns and components of the life cycle for different genotypes in natural populations of the fish *Zoarces*, by Christiansen and Frydenberg (1973). However, the very power of their method depended critically on finding a model species with rather restricted properties: large populations and large broods that could be associated unambiguously with their mothers, that is, live-bearing fish.

In some instances, model studies are an extension of the tradition of natural history, in which the provision of an interesting narrative is an end in itself. An interest in the species for its own sake is certainly an explanation of the large amount of effort devoted to human population genetics. The attempts, all ending in failure, to provide an explanation of human blood group polymorphisms are illuminating. At first sight the model is an ideal one. Biology, history, and social structure are better known for humans than for any other organism, and blood groups are simple genetic polymorphisms. Moreover, in the case of one of these polymorphisms, *Rh*, some important physiological consequences of the phenotype are known. Nevertheless, even the most reduced form of the problematic has not been successful. It is not known whether the pattern of any of these polymorphisms is a consequence of natural selection, not to speak of estimating the fitness of various genotypes. The problem lies in the excessive time and effort that would be needed to acquire complete reproductive and survival information, beginning at birth, for a large enough sample of each genotype.

The alternative to attempting a complete description of the interacting forces responsible for an evolutionary pattern in a model system has been to settle for less detailed explanations, but for cases that seem representative of a large domain of characters and species. Usually this has meant concentrating on one or another causal mechanism, attempting to estimate its magnitude or, at the lowest level, simply testing for whether that magnitude has been sufficient to be effective in the evolution of the characters. It is the clear intent of such programs to produce some generalizations either across genes or across species. The particular case that has preoccupied population geneticists since Kimura's development of the theory of neutral evolution has been distinguishing natural selection from all other causes of evolutionary patterns. The electrophoretic surveys of protein variation, across scores of genes in hundreds of species, that characterized experimental population genetics for twenty years after they were first introduced in 1966 were an attempt to make some statistical generalization about variation and its modulation by natural selection. What proportion of genes show evidence of balancing selection, of strongly constraining purifying selection, of neutral polymorphism? In the last dozen years observations on nucleotide polymorphisms have replaced those on simple protein polymorphism, but with the same intent. Because of the greater cost and time involved in nucleotide sequencing, many fewer genes have been studied, so that estimates of proportions of genes displaying one or another pattern of selection cannot be made with any precision, but phenomena can be seen to be "frequent" or "reasonably common" or "rare."

Whether the genetic variation is detected by protein or nucleotide sequence, studies of individual genes in particular species now can assert a generality that biological model programs cannot. First, the genetic entities studied, nucleotide or amino-acid substitutions without large effects on phenotype, can be assumed in each individual case to be representative of a very large class of natural phenomena. Second, a diversity of such individual cases is being studied in a variety of species so that a communal agreement has been reached that the observed patterns are general. So, for example, Kreitman's (1983) original demonstration from nucleotide sequencing that virtually all amino acid substitutions in alcohol dehydrogenase in *Drosophila melanogaster* are effectively deleterious and are selected out, has since been repeated a number of times with different genes in different species, always with the same result. Indeed, the cumulative evidence is now convincing that no class of nucleotides – not introns, nor redundant coding positions, nor even downstream nontranscribed sequences – are without selective constraints. Third, there are some particular genetic variations that are themselves of intrinsic general interest. All species show an uneven use of alternative equivalent codons for redundantly coded amino acids. These codon biases vary from gene to gene and from species to species. The claim by Akashi (1996) that selection among alternative codons is involved in codon bias in *Drosophila,* and his estimate of an extremely low selection intensity (about 10^{-6}) for the case of the alcohol dehydrogenase gene, are taken to be of general application across genes and species because it is assumed that such selection is insensitive to external environmental conditions or species specific biology, and reflects fundamental cell metabolic demands.

Closest to the analytic program of theoretical population genetics are studies whose problematic is the demonstration that a theoretically possible phenomenon may, in fact, be manifest in a living system. There is no commitment to the frequency in nature with which the phenomenon is effective, but the fact that it can be demonstrated in even one case carries the implication that it cannot be all that rare. An example is the demonstration by Berry, Ajioka, and Kreitman (1991) that a region of the genome of *Drosophila melanogaster* that is completely without recombination is also completely without genetic variation. This result is interpreted as a consequence of a recent rapid sweep through the species of a selectively advantageous gene which, on theoretical grounds, should sweep out all the genetic variation with which it is linked. As it stands, the single observation is only a confirmation of a theoretical possibility that predicts a correlation between the occurrence of regions in the genome of low variation with regions of low recombination. If such an observation could be turned into a frequency pattern by studies of various regions of the genome in various species (or even one species) it would provide an estimate, complementary to the generality of deleterious purifying selection, of how frequently new advantageous mutations become incorporated in a species genome.

MODES OF INFERENCE

The attempts to infer the combination of forces operating on the evolution of heritable traits can be characterized according to two epistemic contrasts. The first, the *functional/tautological,* concerns the way in which the parameters are estimated. The functional alternative, to measure the actual life history parameters of each genotype, is usually not possible except for a few model organisms like *Zoarces.* Even when there are measurable differences in behavior and physiology between genotypes, translating these into fitness is extremely difficult because the differences are likely to be very small in most cases. So, for example, virtually every electrophoretic polymorphism in humans can be shown to be associated with a substantial in vitro measurable difference between genotypes in some aspect of enzyme kinetics. Yet there is no evidence that these differences in kinetics are reflected in differential survivorship or fertility.

For a species whose natural history is not well understood the problem is even more severe. It might be shown that a genetic difference in *Drosophila* results in a substantial difference in physiological fecundity, so that females of one genotype can lay many more eggs than another. Such a fecundity difference would, all other things being equal, represent a large difference in fitness. But all other things are not equal. We do not know whether differential physiological capacity to lay eggs in *Drosophila* contributes anything to the variance of fitness in nature. In fact, it is probably irrelevant for *Drosophila pseudoobscura,* because females lay few eggs at a natural site; the differential ability to find a suitable egg-laying site is probably more important. One cannot assume from functional measurements in laboratory conditions that such differences are translated into fitness differences in nature. Dobzhansky's demonstration that inversions in *Drosophila* undergo repeatable changes in frequency in the laboratory because of natural selection tells us only that under some environmental circumstance there is selection for inversions. In fact, there was selection for inversions in the laboratory at 25° C but not at 18° C, and we know nothing about their selection in nature.

The alternative to making functional measurements is to use theory in a backward direction to provide tautological estimates of forces. The theoretical structure of population genetics is derived by beginning with concrete models of the genetics of traits and of the demography of populations. These models are then transformed into formal mathematical structures with variables and parameters, structures that are used to predict both the kinetics of change of the frequencies of genotypes and the long-term steady states that will occur if the forces operate over sufficiently long periods. Properties of organisms and populations that begin as concrete descriptions of biology, end as parameter values of abstract entities in mathematical formulations. And it is these parameterized mathematical structures that are used to predict concrete genetic properties of populations. There is a biological input and a biological output, connected by a deductive transformation. In this deductive mode, one need only make the appropriate functional measurements of the input process to make a prediction of the output. Suppose, however, that it is not possible to measure by direct functional observation the parameters that characterize the biological input, because they are too small or because the organism is too elusive. One can then attempt to invert the transformation, using the biological result to infer the properties of the causal forces. But what if the original biological model of causation is not an accurate reflection of what is really happening? Then the inferred measurement from the inverse, inductive transformation is not an estimate of the parameters of the original concrete model, but a tautological *als ob* quantity that would have estimated the appropriate parameter had the model been correct in the first place, but which now bears an unknown relationship to the actual biological properties of the population.

The classic example of this tautological inversion in population genetics is the concept of effective population size, N_e. The basic demographic model is that there are $N/2$ males and $N/2$ females (or N hermaphrodites) in a population, that mating occurs between a randomly chosen male and female to produce a single offspring, and that these parents then return to the pool of potential mates. A male and female again couple at random, and this process continues until a new generation of $N/2$ males and $N/2$ females is produced. A demographic consequence of this scheme is that the number of offspring produced by individual parents is Poisson distributed, with roughly one-third of the parents leaving no offspring at all. The genetic consequence of such a mating process is that there is a loss of approximately $1/2N$ of the heterozygosity in the population each generation. In the forward, functional mode, one need only count the population and deduce the genetic consequences. But of course there is no population in the world that matches these specifications, and even if there were, it is not generally possible to count them. Male and female numbers differ, mating is never completely at random, couples leave a variable number of offspring that does not match the Poisson distribution, population numbers do not remain constant from generation to generation, and worst of all, many organisms are continuously breeding over many age classes and so do not even have well-defined generations or a well-defined number of re-producing individuals. What the population geneticist actually does is to make measurements of the rate of decrease of heterozygosity per generation for some genotypes, and then infer an *effective* popu-lation size, N_e, that would account for the observations if the biolog-ical model had been applicable. This entirely tautological estimate of population size has no concrete embodiment in the actual demogra-phy of the population, and there is no conceivable outcome of the genetic observations that would not allow an estimate to be made. Moreover, the result would have been different if different genotypes had been observed, because natural selection has an effect on the distribution of offspring numbers.

The parameters of selection, too, are often estimated tautologically by assuming that observed changes in allele frequency are the conse-quences of a simple scheme of selection, and then back-calculating the selection parameters of the simplest fitness model in which geno-types differ only in their survivorship, but not in their fertilities. Nor are diachronic observations necessary. If it is assumed that the popu-lation is randomly mating, and that the population is at equilibrium, then deviations of genotypic frequencies from Hardy-Weinberg pro-portions can be used to estimate selection coefficients for the sim-plest selection model. But these, again, are tautological.

There are two different issues that arise in tautological inference. First, as we have already discussed, the detailed biology of each process may be quite different from the simple model. Second, the separate tautological estimate of the parameters of one of the forces – say, effective breeding size or selection – assumes that there are no other significant forces operating. In a few instances simultaneous tautological estimates of all the parameters have been attempted, taking into account the full theory of interaction of forces. Lamotte attempted this estimation for the presence or absence of shell ban-ding in *Cepaea* by fitting the observed distribution of frequencies of

the alleles over a large number of populations to a theoretical station-
ary distribution. Given the parameters of population size, selection,
migration, inbreeding, and mutation, and the assumption that the
distribution of allele frequencies over populations is at a stochastic
steady state, it is possible to predict this distribution of allele frequencies. Unfortunately, the parameters appear in the formulation in con-
founded form as the products and sums of parameter values that
cannot be separately estimated, so that the result is extremely unin-
formative biologically.

It is unclear what knowledge is gained from tautological inferences, unless a strong case can be made from independent biological observation that the basic model on which the tautological procedure is based is near enough to correct to allow some concrete
interpretation of the tautological quantity.

The other schematic difference in inference is the *dynamic/static*
distinction. Population genetics aims to provide explanations in
terms of *forces,* pressures that have made the present situation
different from the past and which are operating at present to induce
yet further changes. It might be that present populations are in a
stochastic steady state, but that individual populations are changing
while the distribution of states over populations remains the same.
Even if it is assumed that the forces have produced a rigid stable
equilibrium in a given population, the equilibrium is the result of
compensatory changes within a generation. So-called balancing selection, for example, is not a cancellation of selective forces, but
rather an equilibrium between changes that occur in genotypic frequencies within a generation because of selection, and an exactly
compensatory change that occurs at the beginning of each new generation as a result of Mendelian segregation. It would thus seem
obvious that the appropriate business of population genetics is to
follow the dynamical changes that are occurring within and between
generations, within and between populations. For the most part,
however, that is impossible, because for the most part the dynamical
forces are so weak that their present operation cannot be detected, or
so sporadic that we are unlikely to catch a population in flagrante
delicto. Indeed, there is some contradiction, because if a force, for
example selection causing the increase of a favorable mutation, is
strong enough to make the change observable as a process, it is likely

to be completed while we are not looking! This problem exists even for model systems where genotypes have large fitness differences that can be studied in the laboratory. One of the statistical difficulties of estimating the selective forces operating in an experimental laboratory population is that the population spends very few generations at intermediate allele frequency, where intergenerational changes are large enough to make a reasonably precise estimate of the fitnesses possible.

Functional measures of parameters are one form of dynamic estimation within a single generation. The actual observations of the physical rate of migration, measurements of mutation, and especially direct observation of the effects of different genotypes on life history components of fitness are short-term dynamical measurements. Tautological estimates can also be dynamic if they depend on rates of change of genotypic composition. At one time it had been supposed that dynamical data derived from following gene frequency changes in laboratory populations could reveal the selective forces operating on naturally occurring alleles of enzyme loci. Some early experiments (see, for example, Gibson's 1970 population cage study of alcohol dehydrogenase variants in *Drosophila*) seemed to show reasonably large selection coefficients under conditions that might be expected to be relevant in nature. However, most of these demonstrations turned out to be results of linkages with other, unidentified loci with effects on fitness. When linkages were randomized, the selection disappeared. It is now generally conceded that the selection on polymorphic variation at individual loci must usually be extremely weak, too weak to observe dynamically. We cannot watch a gene in the process of being replaced during speciation, yet we would like to know how much of the genetic differentiation between species occurs by a random as opposed to selective process.

A combination of the weakness of the forces and the lack of any method for observing most individual gene effects on phenotype has resulted in the replacement of dynamic by *static* data as the basis of inference. Even if expected rates of directional genetic change are small, and even if stochastic processes make any particular change uncertain, it should be possible to use the cumulative effect of slow processes over long periods to detect the presence of the weak forces.

That is, the pattern of standing genetic variation within and between populations and species at any moment should be a consequence of the accumulation of very small differences over very long periods. After all, that is the very argument used by Darwin to support his claim for evolution despite the apparent temporal stability of extant species. Moreover, if one is lucky, different patterns of standing variation will be predicted by different combinations of weak evolutionary forces, so that an unambiguous inference can be made.

The use of static data as a basis of inference had already become dominant during the struggle of the 1950s over whether selection on allelic variation was mostly balancing as opposed to purifying. The frequency distribution of homozygous viabilities of whole chromosomes sampled from nature was easy to characterize (in laboratory conditions, of course!). A variety of numerical measures on these distributions were devised (see Lewontin 1974: 68–82, for details) that were supposed to distinguish between hypotheses about selection, but they were unsatisfactory for a number of reasons, chief among them the inevitable confusion between single genes and whole chromosomes.

In the present era of molecular population genetics the reaction to the communal agreement that selection is usually too weak to detect dynamically, and to the challenge made by the neutral theory of evolution to the hegemony of selective arguments for the explanation of genetic variation in space and time, has been to depend almost exclusively on static data as the chief tool in population genetic inference. The method employed is the standard one of statistical inference. We begin with a null model and generate an expected pattern of genetic variation within and between populations which can then be compared to the actual pattern. If the null model is rejected, then the pattern of deviation from expectation may reveal the forces operating. Because the chief interest at present is to detect various forms of selection, the null model is that there is no selection, that the evolution of variation has been entirely the consequence of stochastic fluctuations in the frequency of randomly arising mutations. But like all null hypotheses, the "neutral" model is really a complex null model with large numbers of specifications besides the hypothesis of interest. These may include claims about linkage correlations between genes, about breeding structure and its constancy, about the isolation of the populations from effective immigration, about uniformity of mutation rates, and so on. These other specifications are thrown into the set of "assumptions," but the decision about what part of the complex model is to be included as an "assumption" rather than as a tested "hypothesis" is not simply a matter of subjective interest, but reflects what can be measured. The great strength of electrophoretic and DNA sequencing methods is that they can be applied to any organism irrespective of its biological peculiarities. We do not need model organisms or model traits. They are all equal in the PCR machine. Thus, observations can be extended over a very large domain of otherwise recalcitrant biological systems. But such an extension means that most of the components of the complex null hypothesis, components that are thrown into the "assumption" set, can never be tested to see whether the assumptions about them are justified.

There is one "assumption" that all static data comparisons must make, and that is that the populations are in a stochastic steady state under some set of constant forces. If the populations are not in a steady state, but are continuing to change directionally from their initial condition, then no inferences are possible without knowing the initial condition and the amount of time that has elapsed since the start of the process. Of course, forces do not remain constant over long periods, and there is no "initial condition," but rather an aperiodic series of externally driven perturbations in the history of any population that reset the historical trajectory. Such perturbations and fluctuations in forces would not be important if the relaxation time to the steady state were short, but a short relaxation time implies that the underlying forces are strong, whereas we suppose that they are weak. When we study the pattern of human genetic variation within and between geographic localities we immediately recognize the traces of history, because we have prior knowledge of that history. Cavalli-Sforza, Piazza, and Menozzi (1996) have used clinical patterns of human genotypes to reconstruct prehistoric paths of human migration from Asia and the Middle East into Europe precisely by assuming that the patterns are not in a steady state, but reveal recent historical perturbations. Exactly the same observations if made in, say, *Drosophila* would be interpreted as evidence of some kind of selective gradient, because the steady-state assumption is not up for test. But why should we assume that fruit flies have any less history than people (especially fruit flies that are human commensals)?

The assumption of a steady state amounts to asserting that history does not matter. The reason that this ahistorical assumption is necessary to static comparisons is that historical events have the potential to destroy possibilities of inference from the static data to forces. Virtually all (but not all) observed patterns of standing variation within and between populations can be reconstructed from an historical mixing of several previously isolated populations with different genetic constitutions, followed by an arbitrary period of further evolution of the mixture. If I am allowed to postulate the genetic composition of the contributing populations and the amount of time that has passed since the mixture, and if I am further allowed to postulate an occasional, nonrecurrent selective event, I can explain *almost* any observed pattern.

In practice, population geneticists have a few tests of the null model that they have used repeatedly, faute de mieux. During the period when the information available on genetic variation was restricted to protein variation, the available tests turned out to have too little power to detect most interesting deviations from the null model. The data, which consisted of the frequencies of variant alleles at diallelic or, more rarely, multi-allelic loci, were simply not rich enough in information to discriminate selective from nonselective patterns, and were completely vulnerable to historical contingency. The introduction of DNA sequencing has made it possible to compare the patterns of genetic variation at nucleotide positions with quite different functional properties – synonymous substitutions as opposed to amino acid replacements, for example. These contrasts have, for the first time, allowed some unambiguous inferences from static data, inferences that are not dependent on uncheckable assumptions. Thus, the repeated demonstration in gene after gene that there is much more nucleotide polymorphism for silent changes than for amino acid replacements cannot be interpreted in any other way than that there is a steady and discriminating purifying selection against nearly every amino acid substitution. The large number of polymorphic nucleotide sites and their pattern of assembly into multisite haplotypes has vastly enriched the available data sets. A few new measures making use of this information have been devised to test the neutral null model, and when applied have reasonably often shown a significant difference from the null hypothesis. Unfortunately there is, as yet, too little information about the ability of these tests to discriminate different reasons for rejection. Most important, all such tests depend critically on assuming that the genetic variation is in a steady state so that the ahistorical method of static comparisons can be used.

There is, at least in principle, a great improvement that might be made in the treatment of static data. It might be possible to devise a measure on the genetic structure of a population that had very low statistical power to detect any deviations from the null model except a particular pattern of selection. Even if no single test could be devised with that property, a set of tests with different powers against different alternatives might, when used conjointly, isolate one form of selection as the one responsible for the static pattern observed. It might even be that the conjoint use of the tests would contradict *all* simple selective schemes, leaving nothing but a recent historical perturbation as the remaining alternative. But since it is only *simple* selective schemes that will have been rejected, history can never be detected unambiguously.

We come back, then, to the original problem of explanation in evolutionary genetics that makes it different from other domains. We are trying to explain a unique historical sequence, filled with historical contingency, by reference to a set of weak, interacting, causal forces whose operation is best inferred by making ahistorical assumptions. Under these circumstances the best to which population geneticists can aspire is a formal structure that sets the limits of allowable explanation and a set of existentially modified claims about what has actually happened in the real history of organisms. To demand more is to misunderstand both the nature of the phenomena and the limits of observation.

NOTE

1. When Elliott Sober was once asked what it would take for a biologist to be admitted as a real philosopher, his response was, "Get a paycheck."

REFERENCES

Akashi, H. 1995. "Inferring Weak Selection from Patterns of Polymorphism and Divergence at 'Silent' Sites in Drosophila DNA." *Genetics* 139: 1067–1076.

Berry, A., K. Ajioka, and M. Kreitman. 1991. "Lack of Polymorphism on the Drosophila Fourth Chromosome Resulting from Selection." *Genetics* 129: 1111–1117.

Cavalli-Sforza, L. L., L. Menozzi, and A. Piazza. 1993. *The History and Geography of Human Genes*. Princeton: Princeton University Press.

Christiansen, F. B., O. Frydenberg, and V. Simonsen. 1973. "Selection Component Analysis of Natural Polymorphisms Using Population Samples Including Mother-Child Combinations." *Hereditas* 73: 291–304.

Gibson, J. 1970. "Enzyme Flexibility in *D. melanogaster*." *Nature* 227: 959–960.

Kreitman, M. 1983. "Nucleotide Polymorphism at the Alcohol Dehydrogenase Locus of *D. melanogaster*." *Nature* 304: 412–417.

Lamotte, M. 1951. "Recherches sur la structure génétique des populations naturelles de *Cepaea nemoralis* (L.)." *Bulletin Biologique de France et Belgique*, suppl. 35: 1–238.

Lewontin, R. C. 1974. *The Genetic Basis of Evolutionary Change*. New York: Columbia University Press.

Lewontin, R. C., L. R. Ginzberg and S. D. Tuljapurkar. 1983. "Heterosis as an Explanation for Large Amounts of Polymorphism." *Genetics* 88: 149–170.

Chapter 10

Experimentation in Early Genetics

The Implications of the Historical Character of Science for Scientific Realism

MARGA VICEDO

INTRODUCTION: FROM THEORY TO PRACTICE

Looking back at the last two decades of work in the area of science studies, one can see a shift from an interest in the products of science, its models and theories, to an interest in the processes and practices of science. Traditionally, the philosophy of science primarily aimed to understand the nature of scientific knowledge. To this end, most philosophers concentrated on the study of its products. But recent scholarship argues that if we want to understand science, we must go beyond the clean work of analyzing reconstructed theories and get our hands dirty by examining the practice of science. This view has provided an impetus for studying various aspects of scientific work, such as scientific instruments, research organisms, laboratory procedures, and diverse disciplinary and professional practices.

While there is no doubt that studies on these topics have provided us with a better understanding of the scientific enterprise, one new line of thinking suggests that we should go even further: in order to understand scientific knowledge, philosophers should move from the library to the field, from their books to the workbench. According to some authors, when we look at the process of science, we discover that science is not mainly about theorizing, but about doing. Thus philosophical analyses of the epistemology, methodology, and ontology of science should pay careful attention not only to what scientists say, but also to what they do and how they do it (Kuhn 1962; Cartwright 1983; Giere 1984, 1988; Hacking 1983; Harre 1986; Pickering 1992).

But before we leave our comfortable armchairs to undertake such a dangerous journey, I think we need to reflect a bit further on the relevance of the study of scientific practice for philosophy. In the realm of science, what exactly does *doing* things reveal that *thinking* about things does not? What does *knowing how* tell us about or beyond *knowing that?* In history and philosophy of science seminars and workshops it is now popular to include laboratory sessions or field trips. I have no doubt that these experiences can teach us something. But what can they teach us that is philosophically significant?

Although little work has been done to articulate how analysis of what scientists do can help to solve or dissolve important problems in the philosophy of science, several authors have presented proposals regarding the value of studying scientific practices in one specific area: experimentation. There has been an explosion of work on scientific experiments in the last fifteen years or so, and some philosophers have advanced rather strong claims about the philosophical significance of experimentation (Cartwright 1983; Franklin 1987, 1990; Galison 1987; Gooding et al. 1989; Hacking 1983, 1988a, 1988b, 1988c; Latour and Woolgar 1986; Martinez 1995; Mayo 1996; Shapin and Schaffer 1986). Deborah Mayo recently argued that experimental knowledge "is the key to answering the main philosophical challenges to the objectivity and rationality of science" (Mayo 1996: 13). Mayo is now a member of a group of philosophers that, following Robert Ackerman, she refers to as the New Experimentalists. She includes in this group Robert Ackerman, Nancy Cartwright, Allan Franklin, Peter Galison, Ronald Giere, and Ian Hacking (Mayo 1996, p: 58). According to Mayo, these scholars share the belief that "focusing on aspects of experiments holds the key to solving some problems in the philosophy of science that stem from the tendency to view science 'from theory-dominated' stances" (Mayo 1996: 58).

Here, I want to focus on the relevance of experimentation to epistemology and ontology. The question is: what do experiments have to do with our beliefs about what there is in the world? In many fields, experiments are an important part of scientific methodology. However, it is one thing to know that scientists use certain methods, and it is another to determine the implications of those methods for what we should believe.

Nancy Cartwright and Ian Hacking have argued in different writings that through experiments we learn things about the world and its objects quite independently of our theories about those objects. In their view, we can often be certain of the existence of some unobservable entities and processes even when our theories about them change. In this way, experimentation helps us to maintain a kind of scientific realism – entity realism – that can withstand the criticisms of relativists and antirealists.

In this chapter, I first present the view that experiments are often independent of theory and then present the position of Hacking and Cartwright that "theory-liberated" experimentation can provide the best evidence for scientific realism about unobservable entities. I then analyze a series of experiments from the early history of genetics and review the different historical interpretations of these experiments. Finally, I consider whether Cartwright's and Hacking's views can help us to understand this example. I argue that in light of the historical character of science, any attempt to find a foundation for scientific realism in an "atomistic" conception of experimentation is futile. I conclude that the epistemological separation between experiments and theories is untenable and, therefore, that experimentation by itself cannot be the foundation for entity realism.

THE LIBERATION OF EXPERIMENTATION

Until very recently the few philosophers who paid attention to experimentation focused on its impact for theory construction and evaluation. Some of the New Experimentalists, however, argue that experiments have a value for scientific epistemology and ontology quite independent of theoretical development.

When philosophers tried to understand the nature of scientific knowledge through the analysis of theories, they valued experiments because of their role in testing theoretical claims. For example, Karl Popper gave crucial experiments a central role in the history of science because he thought that they help to corroborate or falsify scientific hypotheses. According to Popper's "epistemological theory of experiment," the theoretician puts specific questions to the experimenter, who then tries to provide definite answers. We could say that in Popper's view, the experimenter is sort of a handmaid to the theoretician, since the latter is the one who defines the problem, formulates the questions, and, in his words, "shows the way" to the experimenter. Furthermore, for Popper, the work of the experimenter is mainly theoretical. Theory, he claimed, "dominates the experimental work from its initial planning up to the finishing touches in the laboratory" (Popper 1961: 107). Experimentation is simply a means of obtaining observations in order to determine the accuracy of a hypothesis. For Popper, then, the epistemological import of experiments lay simply in their relevance to theory testing.

When theory construction (and not only theory justification) became of interest to philosophers, many of them noted that experiments play an important role here as well. In particular, experiments help to uncover phenomena and data that inform the development of theories.

Whatever the role of specific experiments in the history of science, moreover, philosophers have often seen experimentation as being thoroughly embedded in a theoretical framework. Consider the long-standing debate about whether Mendel used his experiments as a means to formulate and develop his theory, to prove it, or simply to illustrate it (see Root-Bernstein 1983 and literature there). In all of the studies on this episode, the question has focused on the role of Mendel's breeding experiments in the formation of his theory of inheritance. In this way, the significance of experimentation has usually been tied to the development of theories. It would therefore seem that experimentation is not interesting on its own. Instead, its value lies in its role in the construction and assessment of theories. From this standpoint, as Sergio Martinez puts it, "experimental activity is subordinated, epistemologically, to the inclusion of its results in theoretical schemes" (Martinez 1995: 4–5; my translation).

Some authors apparently give a more balanced picture of the relationship between theory and experiment, but it is still one in which the ultimate goal is to assess the value of scientific theories. For example, Bas van Fraassen suggests that theory has a twofold role in experimentation: it helps to formulate questions to be answered, and it guides the design of experiments to answer those questions. He notes that experimentation also plays a dual role in the construction of theories: it is used, first, to test the empirical adequacy of a theory and, second, to guide the further construction of a theory. This seems to suggest a somewhat symbiotic relationship between theory and experiment, and he emphasizes the "intimately intertwined development of theory and experimentation" (van Fraassen 1980: 74). Although this position does not give experiments a "subordinate" role in the development of scientific knowledge, experimentation is important here primarily because it tells us which theories are worth developing and accepting. This is so because for van Fraassen, the aim of those scientific activities is "to obtain the empirical information conveyed by the assertion that a theory is or is not empirically adequate" (van Fraassen 1980: 74).

But at least some of the New Experimentalists argue that we need to adopt a perspective less dominated by theory, a stance that values experiments "on their own merits," so to speak. They suggest that as philosophers emphasize practice more and theory less, they will realize that scientific experiments often acquire an increasingly autonomous significance and can eventually become independent of theory. Franklin (1987, 1990) and Mayo (1996) thus propose that there should be an epistemology of experimentation. This does not mean that the epistemology of science must always consider experiments, but rather that we must develop a specific epistemology tailored to experimental practices. Although Hacking says that "the experimentation liberation movement" aims "not just to elaborate the life of experiment, but also to improve the quality of life for theories," (Hacking 1988b: 148) he has also claimed that "experimentation has a life of its own" (Hacking 1983: 150).

But what could it mean to say that experimentation has a life of its own? Mayo argues that this slogan encompasses a series of beliefs that the New Experimentalists share to a greater or lesser degree. The first is the view that experiments aim to discover things not by testing any global theory, but only by testing topical hypotheses. The second is the idea that experimentation can provide a theory-independent warrant for data. The third is the point that experimental knowledge often remains valid after theories have changed (Mayo 1996: 62).

Although these are complex theses that need careful scrutiny, they all suggest that experimental knowledge is sometimes independent of theoretical knowledge and, furthermore, that it is more permanent. Experimental knowledge, by implication, is more "trustworthy" than theoretical knowledge. So we can be more certain of the objectivity and continuity of experimental results in science than of its hypotheses and theories. This suggests, in sum, that experiments are independent of theory and that they allow us access to reality that is superior to the access that scientists obtain through observation and other methods of study.

It may well be the case that experiments can often provide us with a better grasp of reality than theories, observations, and other scientific practices, but is the "experimental grasp" strong enough to allow us to assert the reality of unobservable entities? Comparatively speaking, experiments could sometimes be better methodological and epistemological tools than other scientific approaches. But some of the New Experimentalists make stronger claims; they see experimentation as the new panacea for resolving the long-standing philosophical controversy about scientific realism. Most prominently, Hacking and Cartwright have argued that experiments provide us with good reasons to be realists about "theoretical entities," that is, unobservable entities postulated by some scientific theories. Let us then turn to Cartwright's and Hacking's views about the relationship between experimentation and realism.

THE ARGUMENT FOR ENTITY REALISM BASED ON EXPERIMENTATION

The dispute between scientific realists and antirealists centers upon the question of how certain we can be of our knowledge of the world, specifically in what concerns our theories and theoretical constructs that purport to refer to unobservable entities.

Hacking establishes a separation between entities and theories; he further argues that one can believe in some unobservable entities and processes without believing in the theories in which they are embedded (Hacking 1983: 29). According to him, the question about theories is whether they are true, whereas the question about entities is whether they exist.

Moreover, Hacking argues that experimental work provides the strongest evidence for scientific realism, because reality, in his words, "has more to do with what we do in the world than with what we think about it" (Hacking 1983: 17).

Hacking has graphically described his "instant" conversion to realism about entities which, suggestively, did not happen in his study, but in a laboratory. One day he asked a physicist how one alters the charge on a niobium ball. The physicist responded that one sprays it with positrons to increase the charge or with electrons to decrease the charge. "From that day forth," Hacking tells us, "I've been a scientific realist. So far as I am concerned, if you can spray them then they are real" (Hacking 1983: 23). This doesn't mean that any entity that we experiment with is real. According to Hacking, only when we manipulate an entity in order to experiment on another part of nature can we be sure of its existence (1983: 263).

In Hacking's view, then, manipulating and engineering – not thinking, observing, or theorizing – are the best tools for finding out about reality. Our capacity to use some entities themselves as tools provides the strongest evidence about their reality that we can obtain. Through interventions in nature, we achieve certainty about the existence of some unobservable entities and processes. And most of the time, we achieve this through controlled experiments (Hacking 1983: 262–275).

Nancy Cartwright also defends a type of entity realism, arguing that we have better reasons to believe in some of the entities and processes postulated by modern science than in the laws or theories concerning them. In her book *How The Laws of Physics Lie,* she encourages philosophers to shift their focus from theories to the causal roles of "strange objects" in these theories. She argues that we "can believe in the unexpected entities of quantum electrodynamics if we can give them concrete causal roles; and the rationality of that belief will depend on what experimental evidence supports the exact details of those causal claims" (Cartwright 1983: 8). Furthermore, in her view, "seldom outside of the controlled conditions of an experiment are we in a situation where a cause can legitimately be inferred" (Cartwright 1983: 6). For Cartwright, experimentation, not observation, is the test of existence in physics, because experiments allow us to identify the specific causes of the phenomena under study (Cartwright 1983: 6–7). Thus, again, experimentation seems to hold the epistemological key to accessing reality.

Although we can classify Hacking and Cartwright under the rubric "entity realists" or New Experimentalists, their positions differ in some important respects. For Cartwright, a controlled experiment that allows us to infer the unobservable cause of an observable phenomenon provides sufficient evidence for the claim that the unobservable entity is real. According to Hacking, however, only when we use an entity to affect another part of nature can we legitimately claim that it exists. Thus Cartwright's criterion for giving unobservables a "green card" into the ontological world is much more permissive than Hacking's.

But their positions both make experimentation the main support for scientific realism. They both believe that only controlled experiments allow us to ascertain the existence of unobservable entities and processes. They also share the view that our belief in the existence of those entities is more stable than any belief about their nature, because while theories come and go, entities are often here to stay.

EXPERIMENTATION AND HYPOTHETICAL ENTITIES IN EARLY GENETICS: CASTLE'S HOODED RATS

I now turn to an episode in the early history of genetics in order to help us reflect on these views about experimentation and realism. The field of genetics should be a fertile ground for exploring the ideas of Hacking and Cartwright, because Mendelian genetics postulates that genetic elements – of unknown nature and structure in the early twentieth century – are responsible for the phenotypic patterns obtained through breeding experiments. There are two levels of analysis in Mendelian genetics: one observable – the phenotype – and the other unobservable – the genotype. One of the main goals of genetic studies is to establish reliable inferences from the patterns obtained at the observable level to the unobservable, underlying mechanisms. To do this, one carries out certain crosses of organisms in controlled breeding experiments and then reasons from the phenotypic patterns or the frequency of the distribution of characters to the postulated causes of those patterns. Thus we can explore geneticists' discussions about what they believed about the unobservable causal mechanisms underlying observable regularities.

Specifically, we can see whether the establishment of causal hypotheses or experimental manipulation warranted belief in the existence of the postulated causes. Early American geneticists themselves struggled at great length with this issue. Their work played an important role in clarifying the concept of the gene and in developing the methodology used in breeding experiments for Mendelian studies.

William Ernest Castle designed his experiments with hooded or piebald rats to illuminate questions regarding the role of selection in the creation and preservation of variation. Hooded rats are white rats with a black hood and a thin black stripe down the back. In his early crosses with those rats, Castle was struck by the great variability of the hooded pattern and he decided to test whether selection could modify it. Castle carried out numerous experiments with the collaboration of several students from 1907 until 1919. During this time, he experimented with approximately 50,000 rats.

With his student Hansford MacCurdy, Castle reported in 1907 that the darkly pigmented, uniform coloration pattern characteristic of wild-type rats was dominant over the piebald or "hooded" pattern. A hooded rat crossed with a wild-type rat produced all wild F_1. The F_2 hooded individuals, however, showed a much greater variability regarding the length of the hood and the width of the back stripe that characterizes the hooded type. Since Castle and MacCurdy suspected that hereditary factors produced this variation, they tested to see whether selection could change the hooded pattern (Castle and MacCurdy 1907).

Castle and MacCurdy reported that selection for a decrease in the hooded pattern was effective. After breeding several generations, they had obtained rats with a smaller hood and a thinner back stripe (Castle and MacCurdy 1907). Moreover, as Castle continued these experiments over many generations, he selected rats for both an increase and a decrease in the hooded pattern. Soon he had obtained a set of rats almost completely pigmented and another set almost all white. These results, Castle argued, showed that modifiability of the hooded pattern was a "fact."

Castle argued that these results were contrary to the Mendelian hypothesis of gametic purity. He believed that in cases such as this one, there was no segregation and inheritance did not follow the Mendelian law of segregation. He called this type of inheritance "blending," because the hereditary factors "contaminated" each other or partially blended during meiosis (Castle 1912a: 360).

Castle was aware that there were two possible ways of explaining his experimental results: either the unit-character was itself fluctuating, or there was a stable unit-character whose expression was modified by the summative effect of other unit-characters. Here the unit-character meant the genetic unit or "factor," as scientists usually called it in those days. The first alternative implied that the factor was unstable. The second alternative suggested that the expression of the factor was influenced by other modifying factors (Castle 1911; Castle and Phillips 1914).

By this time, three researchers had already discussed multifactorial inheritance. G. Mendel had addressed the question in 1865. To explain his own results with oats and wheat, the Swedish investigator H. Nilsson-Ehle presented in 1909 the hypothesis that various factors can influence the same phenotypic trait. In the United States, E. M. East, Castle's colleague and the head of plant genetics at Harvard's Bussey Institution, independently arrived in 1910 at the same idea through his studies with corn. He postulated that the effects of factors are cumulative and argued that the multiple-factor hypothesis might account for the inheritance of many quantitative characteristics. In the case of his studies with corn, he suggested that the effects of various factors combined to produce continuous variation, though the genetic factors are discontinuous and their transmission obeys Mendel's law of segregation (Mendel 1865; Nilsson-Ehle 1909; East 1910, East and Hayes 1911).

Castle perceived the debate over the alternative interpretations as a dispute between two camps: the selectionists and the mutationists. The two groups, he argued, held different conceptions of the unit-character. While a selectionist believed in unstable unit-characters, a mutationist conceived of them "as changeless as atoms and as uniform as the capacity of a quart measure" (Castle 1912a: 354).

After considering the multiple-factor hypothesis, however, Castle opted for blending inheritance, arguing for the modifiability of the unit-character. Confident that the genetic factors were unstable and that the power of selection was great, Castle concluded that selection could change the unit-characters. Siding with the selectionists, Castle argued that organisms "are not devoid of variability; neither are the unit-characters which they manifest devoid of variability, nor yet is the germinal basis of such unit-characters devoid of variability. Unit-characters may arise gradually as the result of repeated selection in a particular direction." Consequently, selection "is not a mere agency for the sorting out of unit variations (factors or genes); it is a creative agency by means of which unit characters can be modified and variation can be given a particular direction, the only limits to its action being physiological limits" (Castle 1912b: 278–79).

The first systematic criticism of Castle's experiments with hooded rats came, not surprisingly, from one of the proponents of the multiple-factor hypothesis, East. East argued that the multiple-factor hypothesis could account for Castle's results with the hooded rats. He claimed that multifactorial inheritance allowed geneticists to describe quantitative inheritance in Mendelian terms in the same way that the postulation of hereditary units allowed one to describe qualitative inheritance (East 1912). However, Castle argued that multifactorial Mendelian inheritance explained the phenotypic results only by improperly introducing factors whose presence scientists could not directly deduce from the observable results. He accused East of postulating "merely subjective" entities in order to widen the scope of Mendelism (Castle 1914a: 688). Years before, Castle had already made a plea for realism and simplicity in Mendelian explanations, saying, "let us in no case introduce more factors into our hypotheses than can be shown actually to exist" (Castle 1906: 280).

The basic question, of course, was how one could show the existence of *any* factors. Castle recognized that factors were unobservable entities and that one could only hypothesize about their existence and their behavior. Though he was willing to do this, he did not believe that the introduction of new entities – modifier genes – to explain quantitative inheritance was warranted: "Practically everyone has now abandoned the idea of '*gametic purity*', but the idea of purity has been shifted from the characters which can be seen to vary, to *factors* which may be imagined to be invariable, though they can not be seen." And, he asked, "What ground is there, then, for supposing that in a case where no factors are demonstrable, such factors are *invariable?* This is like supposing that the moon is made of cheese and that further this cheese is *green*" (Castle 1914b: 94, 95). Thus Castle now appealed both to realism and simplicity to argue that one should not introduce more entities than were necessary to explain the observable data. The introduction of modifier genes was not justified, from his point of view.

In fact, both Castle and East appealed to simplicity in order to defend their positions, but, as I have argued elsewhere, simplicity could not have been the decisive factor in this dispute. If one wanted to include modifier genes, one could use Mendelian inheritance to account for the patterns obtained in the cases of qualitative and quantitative inheritance, which meant having only one process and not two (Mendelian and blending inheritance, as Castle's position required). Thus fewer entities or fewer processes was the question, and one could not make this choice based on parsimony alone. Simplicity by itself could not decide which theory was most parsimonious in all respects (Vicedo 1991).

Soon other geneticists joined this debate. H. J. Muller argued in 1914 that his own work with *Drosophila* showed that the "vast majority of genes were extremely constant." And he argued that Castle's results could be explained by postulating the existence of modifier genes. According to Muller, Castle was violating one of the most fundamental principles of genetics – "the non-mixing of factors" – to support another violation of a basic genetic principle – "the constancy of factors" (Muller 1914: 567). Castle responded by questioning the "fundamental" nature of Muller's "constancy of factors": "When did these principles become 'fundamental'; by whom were they established and on what evidence do they rest?" he asked (Castle 1915: 50).

In a review of Castle's work, Raymond Pearl pointed to the basic problem: "the difficulty with all these experiments is not in the facts but in their interpretation" (1917: 88). He went on to argue that selection could not possibly operate *directly* upon germ-cells (Pearl 1917: 74). Thus Pearl did not agree with Castle. In Pearl's view, "the assertion that new variations are caused by selection is the rankest kind of mysticism plus bad logic" (Pearl 1916: 104). He even argued that Castle did not recognize a basic tenet of "genetic epistemology," namely, that one could not use the phenotype as an exact index of the genotype (Pearl 1916: 92). But, of course, this was at the core of the whole issue. At that time, the phenotype was the only guide to the genotype. So the question was this: what type of inferences can one make from the phenotype to the underlying, unobservable causal mechanisms? Pearl argued that selection could not *cause* genetic change. Castle then blamed the differences of interpretation on "the philosophic pitfall of causation" (Castle 1916: 252).

But other important geneticists sided with Castle. Richard Goldschmidt argued that "if . . . it can be proven that genes are substances with the attribute of definite mass, it would be illogical to deny their variability. Nobody will claim that a gene is a substance that passes unaltered from generation to generation." Goldschmidt argued that his own experiments together with Castle's work showed that the "*deus ex machina* modifying factor" was superfluous. Like Castle, he concluded that selection can "change the quantity of the gene, and also, therefore, the somatic characters caused by quantitative differences in the gene, until the physiological limit is reached" (Goldschmidt 1918: 40).

It is not necessary for our purposes to review all the details and arguments of this controversy that lasted over a decade, but it is worth noting that it engaged many of the most important geneticists of the time. It included some of Castle's former students (MacDowell 1916), other members of the *Drosophila* group (Sturtevant 1918), and geneticists from Europe (Hagedoorn and Hagedoorn 1913).

In 1919, Sewall Wright, at the time a student of Castle, suggested an experiment that could help scientists to settle this controversy. The rats from both the plus and minus series were to be crossed with a third, wild race. If the almost pure white and the almost pure black patterns obtained in Castle's experiments had been caused by the accumulation of modifiers, then the crossing with the wild rats would eliminate this difference, and the progeny of successive generations from the two series would eventually be similar.

Eventually, the results of the test proposed by Wright came in: the rats were identical. Castle had tried this procedure before without success. He and Phillips had reported in 1914 that they had crossed both the plus and minus series strains with wild rats and had extracted the hooded pattern again in the F_2 generation. The hooded patterns had regressed a bit toward one another. But since they had still differed significantly, Castle had seen no need to continue this process for more than one cycle. But now, in 1919, Wright suggested following this process for several generations.

After seeing the results, Castle claimed that a "crucial experiment" had settled the question. He accepted the view that selection had not altered the factor or gene responsible for hoodness. Adopting the term "gene" in his papers for the first time (instead of the term "unit-character"), he now admitted that genes were stable and could not be modified by selection. He thus adopted the multiple-factor hypotheses to explain the results previously obtained in crossing hooded rats (Castle 1919a, 1919b).

"World War I had ended; but for Castle it was not Armistice; it was unconditional surrender," according to E. A. Carlson (Carlson 1966: 38). Historians have unanimously agreed that in 1919 Castle had finally resolved his differences with the rest of the genetics community about the interpretation of his results with hooded rats.

But the interesting twist is that if Castle's experiments were difficult to interpret for him and his contemporaries, they have been no less so for historians of science. In the next section I review some of the different readings they have provided and their implications for our understanding of the role of experimentation in science.

WITH HINDSIGHT: CASTLE'S EXPERIMENTS IN THE HISTORICAL LITERATURE

The benefit of hindsight does not seem to have helped historians to reach an agreement on how to interpret Castle's experimental results and on their significance in the history of genetics. First, different scholars do not agree on *what* Castle's experiments "proved." Second, they also disagree about *who* provided the correct interpretation of the results obtained. Third, it remains unclear *how* geneticists reached closure in this controversy. And, fourth, I further ask: what counts as *closure?*

Scholars differ in their assessment of the implications of Castle's results with the hooded rats. For some, Castle's experiments conclusively disproved the views of W. Johannsen about selection. William Provine, for example, argues that Castle's experiments "demonstrated the power of selection" (Provine 1986: 73), and that "more than any other, Castle's experiment forced geneticists to give up the apparent conclusion from de Vries and Johannsen that selection of small differences could lead to no significant genetic modification" (Provine 1986: 72). In a review of early experiments on selection, D. S. Falconer concludes that Castle's experiments disproved "Johannsen's belief that selection merely sorted out existing genotypes." According to Falconer, "Castle here recognized the power of recombination to create new variation" (Falconer 1992: 6). Thus, he concludes, "this experiment very convincingly refuted all the propositions believed by the mutationists" (Falconer 1992: 7). Garland Allen also believes that Castle's work was later vindicated by history. Writing about the hypothesis of "position effect" postulated by T. H. Morgan's group, Allen tells us that "it may seem ironic that those who had so criticized Castle's work less than a decade before should now promote a theory which claimed similar degrees of modifiability of genes" (Allen 1974: 69).

While they agree that Castle's experiments were important, the interpretations provided by those historians vary in important respects. Provine says that Castle's experiments proved that selection could lead to genetic modification. Falconer argues that they showed the power of recombination and the permanency of changes made by selection. Allen believes that Castle's views on the modifiability of the genes were later vindicated by the discovery of position effect. All of these theses are related, but they are not identical. The position effect refers to the expression of genes, not to their stability, and it does not require the modifiability of genes by selection. Provine's and Falconer's theses that selection can lead to genetic modification and that those changes are permanent are rather vague. After all, the whole controversy over Castle's experiments centered around the clarification of those statements, namely, what exactly did it mean that selection can *cause* genetic change? And what did it mean that genes are stable or modifiable entities?

Second, even with the benefit of hindsight, scholars do not agree on *who* interpreted Castle's experiments correctly. Although they offer somewhat different views about the experiments, Provine and Falconer both believe that Castle provided a correct interpretation of his experimental results. Allen also seems to believe that Castle's position was somewhat vindicated by later developments in genetics. More recently, Raphael Falk has argued that "Castle was not wrong in his interpretation of his experimental results; rather, his interpretation was not helpful, not instrumental. It did not allow the expected generalizations to be made" (Falk 1986: 145). All of these views contrast clearly with that of E. A. Carlson, who said that Castle had lost the battle over how to interpret his hooded rat experiments and had "unconditionally surrendered" to the multiple-factor hypothesis proposed by East and others (Carlson 1966: 38).

Neither is there agreement about the factors that helped to "solve or resolve" this controversy. Provine, Falconer, Allen, Carlson, and Cravens (1978: 171) all believe that the debate over the stability of the genetic units was finally closed by empirical evidence obtained through experimentation. For them, the debate ended in 1919, when the results of the experiment proposed by Sewall Wright came in. For Hamilton Cravens, for example, the experiments carried out by Wright and Castle "conclusively demonstrated that Mendelian factors could not be modified, i.e, "blended" (Cravens 1978: 171). Thus closure was achieved through the addition of empirical evidence provided by experimentation. But Raphael Falk offers another interpretation. For Falk, not any particular experiment, but the definition of the term "gene" helped Castle and his contemporaries to clarify the situation. Falk argues that

. . . it was not the experiments with the piebald rats that convinced Castle to give up his resistance: these experiments could easily have been interpreted on his previous assumptions of the modification of units in heterozygotes. The answer given by the Mendelians that resolved the circularity in the argumentation was simple: that trait that 'Mendelizes' (*i.e.* that segregates among the progeny according to the Mendelian laws) is *by definition* determined by a single gene; or, any visible character of an organism which behaves as an indivisible unit of Mendelian inheritance *determines* a unit-character or a gene. (Falk 1986: 143)

So, was it empirical evidence or definition of concepts that provided a "solution" to the controversy? Or maybe both?

Lastly, it is questionable whether the controversy about the experiments with the hooded rats was indeed "resolved" in 1919, as scholars have argued. The historical literature suggests that the experiments performed in 1919 with Sewall Wright provided closure to this debate. Maybe it is true that by this date the debate was "closed" for the scientific community; but it is still necessary to ask why, and to determine what closure meant in this case.

For Castle, the "cruciality" of his 1919 experiments with Wright soon evaporated (Castle and Pincus 1928). Contrary to the standard historical interpretation, he eventually went back to his original view, arguing that while the multiple-factor hypothesis was a possible explanation of "blending inheritance," it had not been proven adequately. As late as 1933 he wrote: "This so-called multiple factor explanation of blending inheritance is ingenious and plausible and, at present, generally accepted among students of genetics, but it lacks demonstrative proof" (1933a: 1012). According to Castle, the multiple-factor hypothesis remained "an hypothesis merely" because its validity had "never been demonstrated by any critical experiment" (Castle 1933b: 184–185).

So we still need further work to assess what factors played a role in "closing" this scientific controversy and to specify what counts as closure for both scientists and historians. In this episode, the main figure on one side of the debate, Castle, changed his mind about multiple factors soon after his "capitulation" in 1919. And the other geneticists were already convinced before 1919 that multifactorial inheritance was the correct interpretation of these results. So, what comprised the "cruciality" of Castle and Wright's experiment? Was this debate truly resolved or was it simply abandoned by the scientific community?

In sum, even with the benefit of hindsight, scholars do not seem to agree about what Castle's experiments showed, who provided the correct interpretation of his experimental results, and how the controversy was resolved. It is interesting to note that several of these scholars – Carlson, Falconer, and Falk – are prominent geneticists. The fact that Carlson believed that the experiments showed that East was correct, that Falconer and Provine argue that Castle was right, and that Falk said that Castle could even "fit" the 1919 results into the position he had defended until then, underscores the malleability of both experimental results and their historical interpretation.

The study of the historical interpretations of Castle's work on hooded rats underscores the malleability of experimental work and shows the need to problematize our notions of closure (Palladino 1996) and consensus (Vicedo 1997). It is clear that we need further work to uncover the complex interrelations among the many social, conceptual, and empirical factors that play a role in bringing an "end" to a scientific controversy (see Engelhardt and Caplan 1987, especially the chapters by E. McMullin, E. Mendelsohn, and R. Giere).

But, here, I mainly want to draw attention to a different point about the historicity of scientific knowledge that will be relevant for our discussion of experimentation and realism in the next section. Both the analysis of Castle's position and the review of the historical interpretations of his work reveal that the way in which we "periodize" our studies can greatly affect what we learn from them. Castle did not see in 1914 that the hooded pattern was reverting to its original form because he stopped his crosses with wild rats after only one generation, probably because the data "fit" his expectations. He ended his experimental series once he had obtained results that supported his other theoretical commitments. Likewise, historical works on Castle typically end in 1919, when he accepted the multiple-factor explanation that we still accept in genetics today. This shows that casting too narrow a net in our research can lead to a distorted picture of the phenomena under study. The analysis of "an experiment" in isolation from its historical development and from the theoretical and experimental traditions in which it was embedded is a misguided practice, one that I believe is too often employed in the history and philosophy of science (Vicedo 1992).

EXPERIMENTATION, REALISM, AND THE HISTORICAL CHARACTER OF SCIENCE

To expand on these points, let me return to the issue of realism and experimentation. The question here is: when we look at Castle's experiments and their interpretations, what can we conclude regarding Hacking's and Cartwright's views on the significance of experimentation for scientific realism? I argue that both positions are flawed because they rely on an unattainable separation of experiment and theory. In addition, the analysis of experiments in isolation from the theoretical and experimental traditions in which they take place prevents one from fully appreciating the historical character of scientific knowledge.

Cartwright, as noted earlier, claims that we can believe in the existence of unobservable entities when we can give them concrete causal roles for which we can provide experimental evidence (1983: 8). With this in mind, consider the question about the cause of the observable pattern in the hooded rats experiments. One could answer it in two different ways, namely, multiple genes or modified genes. Both explanations could provide a causal chain leading from the hypothesized unobservable mechanism to the observable effects.

But the provision of a causal account is not sufficient support for asserting the reality of a hypothesized unobservable entity or process. Having a possible or even plausible causal account is different from having "the real" causal account. Of course, Cartwright does not claim that any causal account provides support for a realist interpretation of hypothetical constructs. According to her, to establish "good" causal inferences, "we reason backwards from the detailed structure of the effects to exactly what characteristics the causes must have in order to bring them about." In addition, we "must have reason to think that this cause, and no other, is the only practical possibility, and it should take a good deal of critical experience to convince us of this." She tells us that the best causal inferences are made in special situations, those in which we can rule out, among other things, the likelihood that other causes are bringing about the effects we are studying (Cartwright 1983: 6).

But to determine the kind of entity or process that could account for the observable patterns under analysis, to rule out the possibility that other entities could bring about those effects, and to discover legitimate candidates for "the most likely cause," one must reason not only from the effects to the cause. One must also consider other realms of phenomena, other data in genetics, and a whole set of cognitive, theoretical, and experimental practices that might help in interpreting the specific effects under study. We need not only to reason "vertically" from the observable level to the unobservable, but also "horizontally" from the observable level to other practices, assumptions, and theories in the field. And, when we do this, we not only draw on experimentation to support our realist commitments; we also bring in theories, assumptions, and hypotheses about the realm of phenomena in which those experiments are embedded. In the case of the hooded rats, what restricted the choice to one explanation or another (modifier genes or modified genes) was the theoretical commitment to Mendelian unification or to the creative power of selection.

Cartwright could say that in this particular example, the most sensible attitude would have been to suspend judgment until further experimental work truly indicated the "most likely cause." But, as we saw earlier, when new experiments were done in 1919 by Castle and Sewall Wright they were insufficient to convince those who were not already convinced – Castle and Goldschmidt. They were also unnecessary to convince those who were already convinced that "the most likely cause" was multifactorial inheritance. What had occurred between 1907 and 1919 was not a refinement of one experiment, an improvement in the techniques used in breeding experiments, or any other experimental advance, but the development of a complex body of theoretical and experimental practices in Mendelian genetics that supported the view that genes are stable entities.

What about the value of Hacking's views in analyzing this example? Hacking, as already noted, tells us that we can believe in the reality of unobservable entities if we can manipulate them to affect other parts of the world. Through specific breeding crosses, Castle manipulated the hereditary material, the factors or genes, to affect another level of nature, the phenotype. Would geneticists at the time have been justified in believing that genes were real because they had been manipulated to obtain certain phenotypic effects?

The problem with answering this question affirmatively is that geneticists held various conceptions of the gene. As we saw, all of those who were involved in the debate accepted the idea that there was a physical basis for heredity. Whether the cause of the phenotypic pattern was "genes" depended on what one meant by genes. If genes were considered to be stable, inmutable entities, then Castle, for example, could not accept their existence. But if genes were considered to be modifiable factors, then he thought that postulating their existence was warranted by the experimental evidence. Our belief that we are manipulating a specific entity will always be intertwined with, and to a certain extent will depend upon, our beliefs about the nature and workings of those entities.

In this case, if one had adopted Hacking's position, one could have said at the time that something was producing the observable results; but this type of realism is so minimal that it cannot be differentiated from many forms of antirealism. After all, most antirealists don't deny that there are entities that we cannot observe and that they produce observable effects. What antirealists deny is that our knowledge about the observable level is sufficient for us to know what those unobservable entities are.

Accepting or denying the existence of genes was tied to discovering the nature and workings of the hereditary factors and on making decisions about how to parcel the underlying level of the genotype and how to conceptualize it. Discovering the existence of genes was completely dependent on empirical information about the degree of discreteness of the hereditary material, about whether or not the expression of that material was affected by other factors, as well as on decisions about how to construct the distinction and the relationship between the phenotype and the genotype, about how to decompose the phenotypic level and the underlying genotypic level into "units," and about how to conceptualize the empirical data obtained. Putting these things together is what we do in elaborating hypotheses and theories about the objects under study. There is no way to intervene "meaningfully" without representing what we are doing, and there is no way to know what we are doing without framing it in a given representational scheme.

Cartwright and Hacking both suggest that only a "minimal" amount of theory is needed to interpret experimental practices. Cartwright allows the use of phenomenological laws, while Hacking allows only the use of what he calls "home truths" (Hacking 1983: 265).

But in the early development of a field, there is no set of shared beliefs. And once there is, we cannot talk about "independent" experiments any longer. By this time, the common practices – both cognitive and experimental – that allow scientists to select from several possible explanations, causal accounts, or beliefs about the manipulation of specific entities are already embedded in a much larger tradition of thinking and doing in that scientific field.

I find it puzzling that Hacking, who has recently been one of the strongest supporters of the existence of scientific styles (Hacking 1992), has at the same time championed the idea that experiments have a life of their own, because one of the most interesting contributions of the idea of scientific styles is that it helps us to focus on the interrelation and integration of different aspects of scientific work (Vicedo 1995). The pioneer writer on scientific styles, Ludwik Fleck, clearly saw this. His views on experimentation are worth quoting at length:

> If a research experiment were well defined, it would be altogether unnecessary to perform it. For the experimental arrangements to be well defined, the outcome must be known in advance; otherwise the procedure cannot be limited and purposeful. The more unknowns there are and the newer a field of research is, the less well defined are the experiments. Once a field has been sufficiently worked over so that the possible conclusions are more or less limited to existence or nonexistence, and perhaps to quantitative determination, the experiments will become increasingly better defined. But they will no longer be independent, because *they are carried along by a system of earlier experiments and decisions,* which is generally the situation in physics and chemistry today. (Fleck 1979: 86)

Fleck argued that a single experiment can hardly carry any conviction. To establish proof, one needs, in his view, an entire system of experiments and controls "set up according to an assumption or style and performed by an expert" (Fleck 1979: 96). Thus only when experiments are embedded in a style of thought, only when "tradition, education, and familiarity have produced *a readiness for stylized (that is, directed and restricted) perception and action*" (Fleck 1979: 84), can we learn through experimentation. Needless to say, I would argue, by this time it does not make sense to talk of experiments as "having a life of their own."

Consider as an illustration Fleck's discussion of the Wassermann reaction and its relation to syphilis. Fleck noted that there is a specific point in time when the causal account becomes a "fact," but that this point "carries" a long history with it. As Fleck saw it:

> *The relation between the Wassermann reaction and syphilis – an undoubted fact – becomes an event in the history of thought.* This fact cannot be proved *with an isolated experiment* but only with broadly based experience; that is, *by a special thought style* built up from earlier knowledge, from many successful and unsuccessful experiments, from much practice and training, and – epistemologically most important – *from several adaptations and transformations of concepts* (Fleck 1979: 97–98).

More recently, Galison (1987) also emphasized the interrelations among the different components of theoretical and experimental work that provide the basis for accepting a piece of evidence as persuasive. Furthermore, he also underscores the "burden" of history: "When experimentalists construct their arguments, the persuasive force of their demonstrations typically depends, at least in part, on theoretical and experimental knowledge imported from the past" (Galison 1987: 6).

Let's now take another look at the case of a scientist spraying electrons in order to alter the charge on the niobium ball. Despite Hacking's conversion story, I doubt that we would immediately convert to realism upon hearing from a physicist that he is spraying electrons, unless we already believed that the long history of theoretical and experimental research in physics was progressing toward a correct understanding of the world and not away from it. If we did not believe that physics is providing us with an approximately correct account of the world, it would probably make little difference what a physicist told us about what he was doing. After all, certain doctrines or theories tell us that we can manipulate forces or other entities (e.g., astrology), but that *per se* does not convince us of their existence. Similarly, experiments by themselves are not sufficient to convince us of the validity of existential claims about unobservables.

In order to assess the "trustworthiness" of an experiment and any associated claims about the existence of unobservable entities, we need to analyze it within the tradition in which it is embedded, as part of an ongoing process of inquiry about a given aspect of the world. As several philosophers of science, including Thomas Nickles (1987) and Dudley Shapere (1982), have emphasized, learning in science consists not only in accumulating knowledge about the world, but in learning how to learn about the world. Experimenting is a crucial part of this process of learning how to learn. In this process, we develop specific rules for doing research in a field, rules that become constitutive of the way of working in a field. These rules are therefore not neutral, for they are essentially intertwined with the substantive empirical claims, theoretical positions, and conceptualizations of the field. Most importantly, the value of specific experimental procedures is intertwined with our views and beliefs about the world in a specific area of science. When successful, this history provides the best reasons for accepting the experimental and theoretical practices of that science. Depending on the success of that tradition, we will have a greater or poorer chance of establishing good causal connections and reliable manipulative practices in specific experiments.

We can now see that entity realism based on experiments "liberated" from theory suffers from an ahistorical view of knowledge. Despite his accusations that philosophers have made a mummy of science (Hacking 1983: 1), Hacking's foundation for scientific realism rests upon a "snapshot" view of history: we manipulate the electron, so now we can believe it exists. The idea that a specific experiment can prove the existence of an entity independently of our theories about it is based on a simplified view of science and its historical character. It slices through a scientific tradition at a specific moment in time by focusing on an experiment in isolation from the long history of cognitive and material practices that have led to the possibility of interpreting it meaningfully.

CONCLUSIONS

In the example I have analyzed from the early history of genetics, we have seen that Castle's experiments "proved" different things to different people. Beliefs that one had a successful causal account about stable or unstable genes, or that one was manipulating multiple stable genes or a single modified gene – such beliefs were continuously revised by both scientists and historians as the theoretical and experimental tradition of Mendelian genetics developed.

This is no different in more mature sciences. In fact, in more established fields the inextricable linkage of our practices and our beliefs is even more acute. As Fleck pointed out, the better defined our experimental practices are, the less independent they are of their theoretical, practical, conceptual, and social frameworks.

These considerations suggest that the position that a specific experiment or series of experiments would confer command for realism independent of the theoretical tradition behind those experiments rests upon a flawed "synchronic" understanding of science that disregards the historical character of knowledge. While trying to "cut nature at its joints," this approach, in fact, artificially cuts into the historical development of experimental and theoretical traditions of knowledge. Instead of trying to privilege experimentation at the epistemological level, a more historically sensitive and philosophically fruitful approach would be the one taken by recent philosophers and historians such as Morrison (1990) and Galison (1987), who try to understand the many ways in which diverse factors and different levels of theory are involved in the design, execution, and interpretation of experiments.

A single experiment or series of experiments cannot command acceptance of the objects the experimenter claims to be manipulating. We do not believe in the existence of "naked entities," objects whose nature we completely ignore. Our beliefs about the existence of an object are necessarily and inextricably intertwined with our beliefs about that object. What we believe we are manipulating depends on our descriptions of and beliefs about the object. So representing is necessarily involved when we conclude that we are manipulating a given entity. Of course, whether or not we manipulate an entity does not depend on our beliefs. But whether we have good reasons to believe that we are manipulating a specific entity does depend on our beliefs about that area of the world and about how we can access it. Thus, experimentation cannot provide a "neutral" or "theory-independent" access to reality. (See also Morrison 1990 for similar criticisms about Hacking's entity realism.)

So, like previous philosophical attempts to provide an overarching, general, neutral rule for deciding which objects we should fully accept in our ontological world, "the experimental argument for realism" (Hacking 1983: 265) is a doomed project. There is no single criterion of existence (inference to a causal pathway or causal efficacy), and there is no one methodology that has epistemological priority (experimentation) in science.

Furthermore, while ontology is independent of epistemology, our knowledge of ontology is not. Hacking tells us that he wants to separate reason from reality because he thinks that "reality has more to do with what we do in the world than with what we think about it" (Hacking 1983: 17). Certainly, reality has little to do with what we think about the world, but the criteria that we use to decide whether we are accessing that reality have a lot to do with how we think about the world and how we think we can operate in it.

ACKNOWLEDGMENTS

Thanks to Jane Maienschein and Rick Creath for inviting me to participate in this project and for their good suggestions about how to improve this article. As usual, Mark Solovey's help has been invaluable. I gratefully acknowledge the support provided by the National Science Foundation and the Arizona State University West SRCA program.

REFERENCES

Achinstein, Peter and Owen Hannaway, eds. 1985. *Observation, Experiment, and Hypothesis in Modern Physical Science.* Cambridge, MA: MIT Press.

Allen, Garland E. 1974. "Opposition to the Mendelian-Chromosome Theory: The Physiological and Developmental Genetics of Richard Goldschmidt." *Journal of the History of Biology* 7: 49–92.

Carlson, Elof Axel. 1966. *The Gene: A Critical History.* Philadelphia: Saunders.

Cartwright, Nancy. 1983. *How the Laws of Physics Lie.* Oxford: Clarendon Press.

Castle, William E. 1906. "Yellow Mice and Gametic Purity." *Science* 24: 275–281.

1911. "The Nature of Unit Characters." *The Harvey Lectures* 6: 90–101.

1912a. "The Inconstancy of Unit-Characters." *American Naturalist* 46: 352–362.

1912b. "Some Biological Principles of Animal Breeding." *The American Breeders' Magazine* 3: 270–282.

1914a. "Multiple Factors in Heredity." *Science* 39: 686–689.

1914b. "Pure Lines and Selection." *Journal of Heredity* 5: 93–97.

1915. "Mr. Muller on the Constancy of Mendelian Factors." *American Naturalist* 49: 37–42.

1916a. "Can Selection Cause Genetic Change?" *American Naturalist* 50: 248–256.

1919a. "Piebald Rats and the Theory of Genes." *Proceedings of the National Academy of Sciences* 5: 126–130.

1919b. "Piebald Rats and Selection, a Correction." *American Naturalist* 53: 370–375.

1933a. "The Gene Theory in Relation to Blending Inheritance." *Proceedings of the Natural Academy of Sciences* 19: 1011–1015.

1933b. "The Incompleteness of Our Knowledge of Heredity in Mammals." *Journal of Mammalogy* 14: 183–188.

Castle, W. E. and Hansford MacCurdy. 1907. "Selection and Cross-breeding in Relation to the Inheritance of Coat-Pigments and Coat-Patterns in Rats and Guinea Pigs." *Carnegie Institution of Washington Publications* No. 70, 50 pp. Washington, D.C.: Carnegie Institution.

Castle, W. E. and J. C. Phillips. 1914. "Piebald Rats and Selection: An Experimental Test of the Effectiveness of Selection and of the Theory of Gametic Purity in Mendelian Crosses." *Carnegie Institution of Washington Publications* No. 195, 54 pp. Washington, D.C.: Carnegie Institution.

Castle, W. E. and Gregory Pincus. 1928. "Hooded Rats and Selection: A Study of the Limitations of the Pure-Line Theory." *The Journal of Experimental Zoology* 50: 409–439.

Cravens, Hamilton. 1978. *The Triumph of Evolution: American Scientists and the Heredity-Environment Controversy, 1900–1941.* Philadelphia: University of Pennsylvania Press.

Diamond, Jared. 1986. "Overview: Laboratory Experiments, Field Experiments, and Natural Experiments." In Jared Diamond and Ted J. Case, eds., *Community Ecology.* New York: Harper & Row.

East, E. M. 1910. "A Mendelian Interpretation of Variation that is Apparently Continuous." *American Naturalist* 44: 65–82.

1912. "The Mendelian Notation as a Description of Physiological Facts." *American Naturalist* 46: 633–655.

East, E. M. and H. K. Hayes. 1911. "Inheritance in Maize." *Conn. Agr. Exp. Sta. Bull.* 167: 1–141.

Engelhardt, H. Tristram and Arthur L. Caplan, eds. 1987. *Scientific Controversies.* Cambridge: Cambridge University Press.

Falconer, D. S. 1992. "Early Selection Experiments." *Annu. Rev. Genetics* 26: 1–14.

Falk, Raphael. 1986. "What Is a Gene?" *Studies in History and Philosophy of Science* 17: 133–173.

Fleck, Ludwik. 1979. *Genesis and Development of a Scientific Fact.* Chicago: University of Chicago Press. Originally published in 1935.

Franklin, Allan. 1987. *The Neglect of Experiment.* Cambridge: Cambridge University Press.

1990. *Experiment, Right or Wrong.* Cambridge: Cambridge University Press.

Galison, Peter. 1987. *How Experiments End.* Chicago: University of Chicago Press.

Giere, Ronald N. 1984. *Understanding Scientific Reasoning,* 2d ed. New York: Holt, Rinehart and Winston.

1987. "Controversies Involving Science and Technology." In Engelhardt and Caplan, eds. *Scientific Controversies,* pp. 125–150.

1988. *Explaining Science.* Chicago: University of Chicago Press.

Goldschmidt, Richard. 1918. "A Preliminary Report on Some Genetic Experiments Concerning Evolution." *American Naturalist* 52: 28–50.

Gooding, David, T. Pinch, and S. Schaffer, eds. 1989. *The Uses of Experiment: Studies of Experimentation in the Natural Sciences.* Cambridge: Cambridge University Press.

Hacking, Ian. 1982. "Experimentation and Scientific Realism." *Philosophical Topics* 13: 71–87.

1983. *Representing and Intervening*. Cambridge: Cambridge University Press.

1988a. "The Participant Irrealist at Large in the Laboratory." *British Journal for the Philosophy of Science* 39: 277–294.

1988b. "Philosophers of Experiment." *PSA 1988* 2: 147–156.

1988c. "On the Stability of Laboratory Science." *The Journal of Philosophy* 85: 507–514.

1992. " 'Style' for Historians and Philosophers." *Studies in the History and Philosophy of Science* 23: 1–20.

Hagedoorn, A. L. and C. Hagedoorn. 1913. "Selection in Pure Lines." *American Breeder's Magazine* 4: 165–168.

Harre, Rom. 1986. *Varieties of Realism*. Oxford: Oxford University Press.

Kuhn, Thomas S. 1962. *The Structure of Scientific Revolutions*. Chicago: University of Chicago Press.

Latour, B. and S. Woolgar. 1986. *Laboratory Life: The Social Construction of Scientific Facts*. Beverly Hills and London: Sage.

MacDowell, E. C. 1916. Piebald Rats and Multiple Factors. *American Naturalist* 50: 719–742.

Maienschein, Jane. 1987. "Arguments for Experimentation in Biology." *PSA 1986* 2: 180–195.

Martinez, Sergio. 1995. "La Autonomia de las Tradiciones Experimentales como Problema Epistemologico." *Critica* 27: 3–48.

Mayo, Deborah G. 1996. *Error and the Growth of Experimental Knowledge*. Chicago: University of Chicago Press.

McMullin, Ernan. 1987. "Scientific Controversy and Its Termination." In Engelhardt and Caplan, eds. *Scientific Controversies*, pp. 49–91.

Mendel, Gregor. 1865. "Versuche uber Pflanzen-Hybriden." *Verhandlungen des naturforschenden Vereines in Brunn* 4: 3–47.

Mendelsohn, Everett. 1987. "The Political Anatomy of Controversy in the Sciences." In Engelhardt and Caplan, eds. *Scientific Controversies*, pp. 93–124.

Morrison, Margaret. 1990. "Theory, Intervention and Realism." *Synthese* 82: 1–22.

Muller, H. J. 1914. "The Bearing of the Selection Experiments of Castle and Phillips on the Variability of Genes." *American Naturalist* 48: 567–576.

Nickles, Thomas. 1987. "Methodology, Heuristics, and Rationality." In J. C. Pitt and M. Pera, eds., *Rational Changes in Science*, 103–1323. Dordrecht: D. Reidel.

1989. "Justification and Experiment." in D. Gooding, et al., The Uses of Experiment, pp. 299–333.

Nilsson-Ehle, Herman. 1909. "Kreuzungsuntersuchungen and Hafer und Weizen." *Lunds Universitets Arsskrift*. N. F. Afd. 2. Bd. 5, Nr. 2, 1–122.

Palladino, Paolo. 1996. "People, Institutions, and Ideas: American and British

Geneticists at the Cold Harbor Symposium on Quantitative Biology, June 1995." *History of Science* 34: 411–450.

Pearl, Raymond. 1916. "Fecundity in the Domestic Fowl and the Selection Problem." *American Naturalist* 50: 89–105.

———. 1917. "The Selection Problem." *American Naturalist* 51: 65–91.

Pickering, A., ed. 1992. *Science as Practice and Culture.* Chicago: University of Chicago Press.

Popper, Karl. 1961. *The Logic of Scientific Discovery.* New York: Science Editions, Inc. First published in 1959.

Provine, William B. 1986. *Sewall Wright and Evolutionary Biology.* Chicago: University of Chicago Press.

Root-Bernstein, R. 1983. "Mendel and Methodology." *History of Science* 21: 275–295.

Shapere, Dudley. 1982. "The Concept of Observation in Science and Philosophy." *Philosophy of Science* 49: 485–525.

Shapin, S. and S. Schaffer. 1986. *Leviathan and the Air Pump: Hobbes, Boyle and the Experimental Life.* Princeton: Princeton University Press.

Sturtevant, A. H. 1918. "An Analysis of the Effects of Selection." *Carnegie Institution of Washington Publications* 264. Washington, D.C.: Carnegie Institution.

van Fraassen, Bas C. 1980. *The Scientific Image.* Oxford: Clarendon Press.

Vicedo, Marga. 1991. "Realism and Simplicity in the Castle-East Debate on the Stability of the Hereditary Units: Rhetorical Devices versus Substantive Methodology." *Studies in History and Philosophy of Science* 22: 201–221.

———. 1992. "Is the History of Science Relevant to the Philosophy of Science?" *PSA 1992* 2: 490–496.

———. 1995. "Scientific Styles: Toward Some Common Ground in the History, Philosophy, and Sociology of Science." *Perspectives on Science* 3: 231–254.

———. 1997. "The Reception of Wilhelm Johannsen's Genotype in American Genetics: Metaphysics, Philology, and Dementia Mendeliana." *Sveriges Utsadesforenings Tidskrift* (Journal of the Swedish Seed Association) 107: 167–177.

Chapter 11

Making Sense of Life

Explanation in Developmental Biology

EVELYN FOX KELLER

Nowhere has the goal of a causal account remained more elusive than in the core question of developmental biology, namely, how does a zygote develop into a multicelled organism? More than any other biological phenomenon, it is the process of embryogenesis that has led so many thinkers of the past to conclude that biology requires distinctive kinds of explanation – in Aristotle's terms, final rather than efficient causes, or in Kant's terms, principles of "self-organization" demanding reciprocity of cause and effect, of parts and wholes. The problem that led Kant to invoke a special "kind of causality" for living systems remains the central problem of development, and it is twofold – namely, the generation in every life cycle of (a) form and (b) function specific to the organism in question. Organisms, he wrote, are the beings that

first afford objective reality to the conception of an end that is an end of nature and not a practical end. They supply natural science with the basis for a teleology . . . that would otherwise be absolutely unjustifiable to introduce into that science – seeing that we are quite unable to perceive a priori the possibility of such a kind of causality. (1790: 558, 566)

As we well know, however, the invocation of goals, ends, or purposes has long posed a major problem for models of scientific explanation. In an effort to establish biology as a legitimate science, more physically minded life scientists of the nineteenth century sought to rid biology of any and all – explicitly or merely implicitly – teleological forms of explanation. By the early part of the twentieth century (particularly in the Anglo-Saxon world) this campaign had for the most part succeeded in effectively banishing all such forms of explanation, and even the very problem of purpose, from respectable life science. Their death sentence was to be subsumed under the label of "vitalism" – a term that became progressively more damning in the early decades of this century.

Since then, and largely as a result of twentieth-century developments in genetics and molecular biology – both disciplines that are committed to the goals of causal-mechanistic explanation – biology has become a science on a par with physics. The long-sought efficient causality came in two successive guises: first, in the earlier part of the century, with the identification of genes (coupled with the causal properties attributed to them in prevailing notions of "gene action"), and later, in the middle part of the century, with the more definitive identification of DNA as the carrier of genetic information (coupled with the notion of a "genetic program").

To most developmental biologists, however, it remained evident that genetics, coupled either with the notion of gene action or with that of a genetic program, faced a serious problem in explaining development – most conspicuously, in the apparent paradox deriving on the one hand from the assumed genetic identity of all cells in a complex organism (and hence the identity of their genetic programs), and on the other hand, from the inescapable demands of developmental differentiation.[1] Nonetheless, in 1970, François Jacob was able to claim, and with remarkable persuasiveness, that "when heredity is described as a coded programme in a sequence of chemical radicals, the paradox [of development] disappears" (p. 4). The organism, to Jacob, was "the realization of a programme prescribed by its heredity" (p. 2). Furthermore, he argued that the genetic program, written in the alphabet of nucleotides, is what is responsible for the apparent purposiveness of biological development. Referring to the often quoted characterization of teleology as a mistress whom biologists "could not do without, but did not care to be seen with in public," he writes, "The concept of programme has made an honest woman of teleology" (pp. 8–9).

Jacob was not the first, nor was he the last, to proclaim a solution to the paradox of development; indeed, one might say that premature announcements of the arrival of an explanation of development have been a constant in the history of genetics. Today, with the extraordinarily dramatic developments of the last twenty years in this

area, most researchers would admit that Jacob too had been premature. Not a few would maintain that it is only today that we can
rightly claim to have arrived at an understanding of the "basic principles" (Wolpert 1994: 571).

Others, however, see things differently. Adam Wilkins, for example, writes that,

> The added wealth of molecular description in the last two decades has
> certainly deepened our understanding, but the underlying mechanisms still
> remain largely hidden. Even the constellation of ideas about progressive
> changes in gene expression, which is the closest approximation we have to a
> general explanation of the hidden dynamics of development, lacks for the
> most part predictive power; it has cogency as . . . a framework to think about
> development, but it is far from a theory of biological development. The
> fundamental problems of understanding developmental change are, there-
> fore, to a large extent still very much with us . . . (1993: 16)

Indeed, one might even say that, for many, it is the very success of
molecular developmental genetics that has brought a new recognition of limits to the explanatory force of genes (or of genetic programs) in developmental biology – or that is what I infer from the
number of critical commentaries appearing in the recent literature of
molecular biology (see, e.g., Brenner et al. 1990; Tautz 1992; Strohmann 1995).

In this chapter, I will examine the uses and limitations of the
notion of genes as causes of (or instructions for) development, and
juxtapose this (genetic) paradigm with an alternative perspective,
namely one that takes as its starting point not genes, but rather the
manifest robustness of the developmental process. These two perspectives lead to radically different questions, and to radically
different criteria of explanation. The role of redundancy in these
frameworks is illustrative of the differences: in one, redundancy appears as an unexpected and possibly inexplicable problem; in the
other, redundancy plays a central role in establishing robustness, as a
first and necessary ingredient of an acceptable explanation.

FIRST, GENE ACTION

Elsewhere (1995), I have discussed the rise of a "discourse of gene
action" among American geneticists in the 1920s and 1930s as a way

of accounting for development. Certainly, little was known at the time about what genes are or how they work, but the discourse of gene action provided a research agenda that could sidestep these lacunae. But in order for genes to count as an explanation of development, it was necessary first to reformulate the traditional problem of embryology. Alfred Sturtevant provided an explicit translation of that problem, and in doing so, spelled out a concrete research program to be conducted with the tools of genetics. He wrote:

One of the central problems of biology is that of differentiation – how does an egg develop into a complex many-celled organism? That is, of course, the traditional major problem of embryology; but it also appears in genetics in the form of the question, "How do genes produce their effects?" (1932: 304)

Between "the direct activity of a gene and the end product," he went on to argue, "is a chain of reaction." The task of the geneticist is to analyze these "chains of reaction into their individual links."

But the discourse of gene action provided more than a research agenda; it also satisfied a strong narrative need. Part of that narrative satisfaction lay in the promise of future control. Here, the precedent provided by physics was crucial, as evidenced by the frequency with which it was invoked. In effect, genetics was to provide the analogue of the atomic theory of matter, with genes as the fundamental units of life. H. J. Muller was perhaps the most explicit in making this argument when, in 1916, he first articulated a parallel he never subsequently lost sight of: the parallel between the problem of mutation, "a control of [which] might obviously place the process of evolution in our hands," and "that of the transmutation of the elements," where

. . . if a means were found of influencing it, we might have inanimate matter practically at our disposal, and would hold the key to unthinkable stores of concentrated energy that would render possible any achievement with inanimate things. Mutation and Transmutation [are] the two keystones of our rainbow bridges to power. (quoted in Carlson 1971; see also Keller 1990)

Much has happened since those early days, in biology as elsewhere. As a result of the extraordinary developments in molecular biology, we now have a vast amount of information about the nature of genes and about their involvement in specific developmental phenomena. In conjunction with these developments, "gene action" has (for the most part) given way to "gene activation." Nevertheless, crucial elements of that earlier discourse persist, most notably in the widespread notion that the development of a trait or function has been *explained* when the gene or genes "for" that trait or function have been identified, and correlatively, in the notion that development as a whole can be explained by enumerating the genes responsible. Indeed, this notion underlies not only genetic explanations of development, but also the very logic of much if not most of genetic research.

But what exactly is the meaning of such claims, that is, claims of the form "gene x is *for* trait (or process, or function) A"? Certainly not that gene x is sufficient for the development of A, since many other factors (including mutations in other genes) may prove capable of preventing or disrupting the development of A. Nor can the claim imply that gene x is involved *only* in the development of A, for x may be (and indeed usually is) involved in many other processes as well. Rather, what is implied by the claim is the presumption – based on the observation that a mutation in x prevents or disrupts the development of A – that x is *necessary* for A. Current research has, however, provided numerous examples in which a second mutation, in a different gene, say y, is found to restore A, showing clearly that claims that x is necessary for A must be treated as both provisional and context-dependent. As Miklus and Rubin conclude, "A gene knockout [or mutation] can result in different phenotypes when it is placed in different genetic backgrounds" (1996: 524). Qualifications commonly appearing in the literature – such as, gene x "is usually necessary for," "is involved in," or more simply "has something to do with" the appearance of A – tacitly ackowledge this limitation, thereby enabling the research to continue, but such qualifications obviously weaken the explanatory force of gene x as a cause.

A complementary challenge to the genetic paradigm arises from what might be thought of as the obverse phenomenon, namely the appearance of null mutants which show no phenotypic effect even when both alleles of the gene in question are dysfunctional. These mutants are called "null" because of their failure to produce particular enzymes that are produced by the wild type, and that are assumed to serve an important function because of the extent to which they have been conserved over evolution. Such "null" mutants came as a big surprise when they were first identified in the early 1970s, and they were generally regarded as quite mystifying. Over the last twenty years, they have become quite commonplace; but they are still seen as somewhat mystifying.

Thanks largely to the acquisition of new molecular techniques for targeted mutagenesis, evidence for such phenomena has accumulated exponentially, and the consensus among researchers in the field is that they clearly indicate the existence of functionally redundant genetic pathways. Today, functional redundancy – on the level of genetic transcription, transcriptional activation, genetic pathways, and intercellular interactions – has emerged as a prominent feature of developmental organization in complex organisms.

Redundancy is a staple of engineering, but it has been said to "strike fear in the heart of geneticists" (Brenner et al. 1991), and for good reason: it not only reveals limits to the value of mutation screening (the core technique of genetics) in probing developmental dynamics, limits that had earlier been only implicit;[2] it is also seen as a threat to the entire explanatory framework of the genetic paradigm. As Diethard Tautz writes,

Though the geneticist will often be unable to say exactly how a certain mutation causes a certain phenotype, . . . he must maintain that single and direct causal relationships exist. This genetic paradigm is at the basis of all systematic mutagenesis experiments, which aim to obtain particular phenotypes, since these experiments usually allow one to look only at the effects of a single mutation at a time. . . . [But] even the best paradigm eventually meets a crisis. Such a crisis is imminent. (1992: 263)

These two challenges to the assumption of genes as the causal units of development – one a consequence of inhibition (or repression), the other of redundancy – are both features of the complex regulatory networks in which genes are now found to be enmeshed. Here is how the Nobel laureate J. Michael Bishop describes the problem:

What at first appeared as simple linear arrays of switches has now emerged as an elaborate network with hundreds (if not thousands) of nodal points. Attempts to trace a signal through the circuitry soon become lost in a welter of crosstalk and feedback. . . . It seems unlikely that [our] efforts will be fully

249

effective unless a global view of the molecular circuitry can be achieved. (1995: 1617)[3]

What would such a global view entail? What kind of explanation might it be expected to yield? And finally what, if any, alternatives to the present explanatory framework of genetics might be available for furthering our thinking about development?

EPIGENESIS AND EPIGENETIC PATHWAYS

The robustness of development, understood in the most general terms as the capacity of developing organisms to compensate for disruptions, has been a primary concern for embryologists ever since Driesch's initial observation of regulation (1891). In an effort to explain instances of robustness at later stages of development, a specific reference to engineering principles quickly surfaced (see, e.g., Spemann 1907) – the term "double assurance." Spemann elaborated on this "principle" in his later Silliman Lectures, referring to it as a "synergetical principle of development." He writes,

The expression "double assurance" is an engineering term. The cautious engineer makes a construction so strong and durable that it will be able to stand a load which in practice it will never have to bear. (1938: 92–3)

But it was C. H. Waddington who, in the 1940s, first attempted to relate observations of developmental stability to genetics, and to outline an explanatory framework in terms of interacting "gene-protein systems." Waddington coined the term *creode* to denote the developmental pathway, or trajectory, that resulted from the interaction of organized genetic systems and environmental effects, and the term "developmental canalization" to denote the progressive stability of these trajectories. The notion of canalization drew both from Whitehead (see Gilbert 1991: 199) and from Waddington's own earlier and explicitly metaphorical image of an epigenetic landscape: a terrain of branching and progressively deeper valleys representing the increasingly stable development of cells toward their adult state. In other words, his particular concern here was with the stabilization of cell differentiation, rather than with the overall robustness of embryogenesis; indeed, his metaphor of alternative pathways breaks down for embryogenesis as a whole, where the key point is the (relative) fixedness of the final state a zygote will reach. Nevertheless, his picture of the epigenetic landscape was useful in drawing his attention to stabilized or buffered pathways of change as a central feature of development (see, e.g., Waddington 1957). An explanation of such developmental canalization, he argued, required supplementing the "genetical theory of genes" with an "epigenetic theory," replacing the standard "atomic theory" with a "quasi-atomic theory" – one in which discrete and separate entities of classical genetics would be displaced by collections of genes that could "lock in" development through their interactions (Waddington 1948).

Waddington's focus on developmental stability guided both his experimental and theoretical efforts throughout the forties and fifties. In one set of studies, by selecting for mutants exhibiting greater than usual variability in pattern formation (e.g., in the number of scutellum bristles in *Drosophila*), and then subjecting these to intense selection pressures, he was able to demonstrate the existence of considerable genetic variability for bristle number in the wild-type population. The fact that this variability is not expressed in wild-type individuals shows, he argued, "that in them the normal four-bristle pattern is in some way stabilized or buffered. The effect of the abnormal gene . . . must have been to produce some destabilization of the pattern, so that the inherent genetic variability could come to expression and be submitted to selection" (1962: 226).[4] In other studies, focused on observed variations in expressivity or penetrance of particular genes, he was able to demonstrate the presence of other (background) genes affecting the expression of the genes in question (see, e.g., 1954). These observations led him, in turn, to the formulation of theoretical models of interacting genes and proteins as a way of accounting for such stabilizing mechanisms.

Today, Waddington seems to be enjoying something of a revival, and it has been suggested that his work on canalization constitutes "a premature discovery" (Wilkins 1997: 257). But in his own time it held little interest for most of his colleagues, and in the decades that followed it scarcely influenced the course of developmental research.

Why is that? One might have thought that the stability (or fidelity) of the developmental process is its most conspicuous feature; indeed, its most basic feature, even the essential precondition of evolution, for without an internal capacity to withstand the inevitable vicissitudes of ordinary development, organisms of a particular genotype would not persist long enough for selection to act upon them.[5] It is precisely the remarkable constancy of the developmental process – its resilience in the face of so much cytoplasmic, environmental, and even genetic variation – that had led earlier thinkers to think of it as inherently goal-oriented, as internally directed toward a fixed final state. And, as already noted, it was this same feature of constancy that the notion of a genetic program was intended to capture. But however the notion of a genetic program is to be construed, the only account for stability it was ever able to provide was limited to whatever structural stability could be attributed to the DNA itself, coupled with the degree of fidelity that could be attributed to its capacity for spontaneous "self-replication" and transmission of "instructions." What that accounts for, as it turns out, is not much. The structural stability of the chromosome is not provided by the DNA itself but by the proteins with which it interacts, and the degree of fidelity one actually observes in both copying and transmission is now known to depend on the presence of an elaborate machinery of proofreading and repair.[6] In fact, the constancy of development depends on its overall stability, and that striking feature remains effectively unaccounted for.

The most interesting issue to me, however, is that a focus on developmental stability points to questions (and hence to explanations) radically different from any than can even be asked within the genetic paradigm. In effect, by its dependence on mutational analysis, the latter seeks to explain development by asking what causes it to fail (or go astray); the assumption is that the causes of normal development can be inferred by logical subtraction – that is, by an enumeration of all those genes that can be identified by their phenotypic failure. By contrast, a focus on developmental stability leads one to ask, what is required to make it work, what is it that endows the developmental process with such reliability? This distinction is familiar to any engineer attempting to formulate design principles for complex systems predicated on the reliability of performance (e.g., the design of airplanes reasonably certain to reach their destinations despite vicissitudes of weather, air traffic, etc.). Engineers have long understood the crucial importance of redundancy in guaranteeing reliability. In other words, their aims (or concerns) led them early on to just that feature of design that has been most inaccessible to the traditional techniques of genetic analysis.

Perhaps, then, it is no surprise that evidence for widespread redundancy in developmental systems has prompted talk of a crisis in the traditional paradigm of genetics. From that perspective, redundancy is not only technically opaque, but it also doesn't seem to make a lot of sense. If genes are assumed to be the units of selection, how could redundancy have evolved? As J. H. Thomas puts it,

> It is perhaps surprising that redundancy is so prevalent, since it is not immediately obvious what selective advantage it might confer. Possession of two fully redundant genes should, on evolutionary time scales, be an unstable condition. . . . [A] similar argument might suggest that even partially redundant genes would tend to lose their redundancy. (1993: 395)

Indeed, it is just this need to make evolutionary sense of redundancy that leads Diethard Tautz to invoke a familiar lesson from information theory – namely, that fidelity in the transmission of information requires redundancy – and to suggest an obvious analogue for living systems. He writes,

> The formation of an adult organism can be seen as the transmission of information which is laid down in the egg and its genome. . . . At each [developmental step] there is a potential loss of information and the developing organism has to safeguard itself against this loss. This is, of course, a good basis for selection pressure to evolve redundancies. This selection pressure need not be very high, since even a small effect on the probability of successful completion of embryogenesis would directly be reflected in the probability of survival of the offspring. . . . Thus, the evolution of redundant regulatory pathways may be seen as a logical consequence of the evolution of complex metazoan life. (1992: 264).

Or perhaps as the logical precondition of metazoan life. Either way, the significant point is that the need to make sense of redundancy has led Tautz and others to just the kind of global view that Bishop advocates. Here, the unit of selection is not the gene, but the life cycle itself.

Questions addressing the stability of the developmental process, or of the life cycle, clearly demand different kinds of tools (both experimental and analytical) than those focused on the effects of mutated genes; but such tools are, at least in principle, readily accessible in the larger world of science. (I am thinking, e.g., of tools from mathematics and computer science that are widely used to analyze the dynamics and stability of biochemical or genetic networks, as well as of the kinds of tools that engineers have long employed to understand – and to guarantee – the reliability of complex systems.)

Waddington had in fact sought to employ at least some of these tools, but hindsight shows us that his approach was conspicuously less successful in fostering productive research than were models of gene action. But does hindsight also show us why? A common assumption might be that, given the state of the art at the time, his questions were not experimentally researchable. And it is certainly true that he lacked the tools of contemporary molecular biology. But he did have the tools of genetics that were then available, and his papers show that he was able to put those to productive use in his own research. The fact that his efforts did not inspire his contemporaries to follow them does not tell us that his questions were not "researchable," but only that they were not pursued. Most importantly, it does not tell us why they were not. Many different reasons might account for the failure of Waddington's program – reasons having to do as much with funding opportunities, scientific fads, and culturally informed explanatory commitments as with the technical opportunities then available. Indeed, all of these various reasons bear on the de facto "doability" of Waddington's program, and only close historical analysis would enable us to distinguish among them, or to weigh the importance of their respective roles.

It might be helpful to such historical analysis, however, to identify some of the different kinds of functions that "explanations" serve in scientific research, above and beyond whatever logical satisfaction they might provide. So, in conclusion, let me very briefly suggest some considerations that I believe need to be taken into account.

DISCUSSION

I have already referred to the utility of "explanations" in guiding research, that is, in suggesting "doable" experiments. But this function of doability needs unpacking: experiments are doable to the extent that the materials, skills, and tools they require are readily at hand (i.e., these need to be more or less "on the shelf"); to the extent that they appeal to the interests of the immediate disciplinary community; and to the extent that they will be funded (through either private or public sources). The meeting of each of these conditions is of course a complex matter, depending, as recent historical and social analyses have clearly shown, on factors well beyond sheer logical and technical possibility. And one of the factors on which they depend is the reigning explanatory framework.

Explanations can often function quite directly (in what I might call their narrative function) to appeal to the interests of the community, unself-consciously invoked to frame (and undergird) an experimental procedure that may be in considerable tension with, if not in direct contradiction to, the explanation (or narrative) proffered. (This is the case, for example, in the technical – i.e., "in-house" – descriptions of the production of "Dolly" – see Keller and Ahouse 1997.) And, of course, explanations are also employed (usually with far more self-consciousness) to satisfy funding interests, that is, to raise financial support for research. This last is not a trivial point, and it is crucial for understanding the widespread hyperbole that accompanies so many of the claims that have been made on behalf of the Human Genome Project and, more specifically, on behalf of the explanatory power of sequence information. Take, for example, a prediction by Harvey Lodish appearing in the same article (in *Science*) as Bishop's:

Eventually the DNA sequence base will be expanded to cover genes important for traits such as speech and musical ability; [when transferred to a supercomputer] the mother will be able to hear the embryo – as an adult – speak or sing. (*Science* 267: 1609)

Such a prediction is of course fanciful, without even the possibility of realization. But the identification of genes and gene sequences has been proven to have enormous predictive and even constructive value, which brings me to what may well be the most conspicuous function of explanation in contemporary developmental biology, namely, its practical or "how to" function.

Notwithstanding the many logical difficulties that beset the notion of genes as causes, the use value of genes as "handles" for bringing about specific effects[7] in both laboratory and industrial settings is undeniable. And for many molecular biologists, "use value" is the goal (and perhaps even the test)[8] of an explanation. Here it is assumed that a proper answer to a "why" or a "how" question must at the same time also be an answer to a "how to" question: an explanation should ideally provide a recipe for construction; at the very least, it should provide us with effective means of intervening. Indeed, Robert Weinberg tells us they are all the same: the reason molecular biologists are convinced that genes are the causal agents of development, and that "the invisible agents they study can explain . . . the complexity of life," he writes, is that, by manipulating these agents, it is now "possible to change critical elements of the biological blueprint at will" (1985: 48, quoted in Woodward). And in a similar vein, Elliott Meyerowitz writes:

Equipped with a predictive genetic model and a knowledge of some of the genes that specify floral organs, researchers now have a considerable degree of control over the development of flowers . . . [providing] living testaments to the power that our growing understanding of the molecular basis of flower development already grants us over the structure and function of some plants. (1994)

But even here, molecular biologists have had to recognize definite limits to control, at least insofar as they are guided by current theory. Sydney Brenner makes the point well in his criticism of the operon model: "The [operon] paradigm does not tell us how to make a mouse but only how to make a switch" (1990).[9]

Contra the view most familiar to philosophers of science (see, e.g., Kitcher and Salmon 1989) – namely, that explanatory adequacy (and/or explanatory power) is self-evident in science, and that "inference to the best explanation" can therefore account for rational theory choice – I would argue that, in actual practice, what counts as an explanation in scientific work is far from self-evident. Here, by focusing on developmental biology, I have tried to show that explanations can function to meet a range of different needs, not only in everyday life but also in scientific practice. One might classify such needs in various ways: for example, prediction, control, and narrative satisfaction is one possible taxonomy. Alternatively, one might think in terms of cognitive, instrumental, and social/psychological needs. Typically, explanations satisfy all of these needs at once, to one degree or another; it is the relative importance of each need that varies. Analysis of the biological literature makes abundantly clear just how much variation exists – over time and between different research schools – regarding the criteria by which an explanation is to be judged. Such analysis also reveals considerable flexibility in criteria, suggesting a far more problematic role for explanation than is often assumed. It would seem that explanations function more locally than is usually recognized, meeting needs determined by the context of particular experimental systems; and yet, in another sense, more globally – in the sense, that is, of meeting needs of social and even political communities extending far beyond the laboratory.

The discourse of gene action, for example, provided a narrative of considerable utility in the context of classical genetics; but it also met other kinds of needs. As I have argued elsewhere, it provided the community of American geneticists during the interwar period with important rhetorical leverage in at least three different domains of contestation: namely, in interdisciplinary politics, international politics, and the politics of gender. Waddington's epigenetic pathways were inspired, in part, by the Whiteheadian philosophy (and its emphasis on process) that was so influential to participants in the Theoretical Biology Club of the 1930s, and by his chosen status of trader between genetics and embryology. In turn, the reception of his work on canalization and "genetic assimilation" was surely influenced by its association with the arguments of Lysenko, an association that was especially problematic during the cold war era (Gilbert 1991: 205).

But just as the meanings and functions of explanation are both local and global, contingent and contextual, we might think in the same way of the very character of the explanatory quest – that is, of what it means to understand, in the most strictly cognitive terms. This would seem to be especially appropriate in developmental biology, where what is to be understood exhibits the same order of complexity as the world of human interests out of which any scientific endeavor necessarily arises.

NOTES

1. For reviews of the conflicts between genetics and developmental biology, see Burian (1996), Gilbert (1992), and Keller (1995).
2. Such limits were inferable, e.g., from the ubiquitous phenomenon of variable penetrance (see, e.g., Wilkins 1997: 213).
3. Moss (1992) makes the point yet more strongly:

> Attempts at securing a discrete causal account of even the most proximate events in the transcriptional activation of a certain gene can thus be seen to quickly devolve into an array of antecedent conditions which are exponentially more complex than the event one was trying to account for. Explorations of the mechanisms involved at the level of the DNA molecule itself, have not led to any privileged point of causal origins, but rather immediately refer back to the complex state of the cell/organism as a whole as the causal basis of the activity of the genes.

4. He continues, "The process is very similar to that in which an environmental stress is utilized to destabilize a developmental system and to reveal genetic variation which was previously concealed. Selection on this variation may eventually lead to the genetic assimilation of the phenotypic modification produced by the environmental stress." (Waddington 1962: 226–7).
5. This view suggests that the role of contingency in development is the obverse of its role in evolution – where contingency is said to be the definitive characteristic of evolution, it is resistance to contingency that could be said to be definitive of development.
6. As Alberts et al. write, "genetic information can be stored stably in DNA sequences only because a large variety of DNA repair enzymes continuously scan the DNA and replace the damaged molecules." (1989; 227)
7. Collingwood long ago argued that this is the sense in which the term *cause* is used in the "practical sciences" (1940: 296–312); and Jim Woodward has more recently emphasized the importance of reexamining this dimension of scientific explanation in the context of contemporary philosophical discussions (forthcoming).
8. As indicated by Phil Sharp's response to my query about the status of explanation in developmental biology: "We will know we have an explanation of development when we can make it happen in the lab" (private communication, 1997).
9. By contrast, Wilmut and his colleagues *have* provided us with a recipe for making a sheep, but without any pretense of "understanding" why their technique should work. Furthermore, complete cells were the object of their manipulation, rather than genes per se. Thus, explanations of their success in terms of a "genetic program" would have to be seen as func-

tioning rhetorically rather than instrumentally (see Keller and Ahouse 1997).

REFERENCES

Alberts, B., D. Bray, J. Lewis, M. Raff, K. Roberts, and J. D. Watson. 1989. *The Molecular Biology of the Cell*. New York: Garland Press.

Beatty, John. 1995. "The Evolutionary Contingency Thesis." In G. Wolters and J. G. Lennox, eds., *Concepts, Theories, and Rationality in the Biological Sciences*, 45–81. Pittsburgh: University of Pittsburgh Press.

Bishop, J. Michael. 1995. Contribution to "Through the Glass Lightly" section. *Science* 267: 1617.

Brenner, Sydney, William Dove, Ira Herskowitz, and Rene Thomas. 1990. "Genes and Development: Molecular and Logical Themes." *Genetics* 126: 479–486.

Burian, Richard M. 1996. "On Conflicts between Genetic and Developmental Viewpoints – and their Resolution in Molecular Biology." In K. Doets, ed., *Invited Papers for the Tenth International Congress of Logic, Methodology, and Philosophy of Science*. Dordrecht: Kluwer.

Carlson, Elof. 1971. "An Unacknowledged Founding of Molecular Biology: H. J. Muller's Contributions to Gene Theory, 1910–1936." *J. Hist. Biol.* 4(1): 160–161.

Collingwood, R. G. 1940. *An Essay on Metaphysics*. Oxford: Clarendon Press.

Darwin, Charles. 1862. *On the Various Contrivances by which British and Foreign Orchids are Fertilised by Insects, and on the Good Effects of Intercrossing*. London: John Murray.

Delbruck, Max. 1966 [1948]. "A Physicist Looks at Biology.' In John Cairns, G. S. Stent, and J. D. Watson, *Phage and the Origins of Molecular Biology*, 9–22. Cold Spring Harbor: Cold Spring Harbor Laboratory Press.

Gilbert, Scott. 1991. "Induction and the Origins of Developmental Genetics." in S. Gilbert, ed., *A Conceptual History of Modern Embryology*, 181–205. Baltimore: Johns Hopkins University Press.

Jacob, François. 1970. *The Logic of Life*. English paperback edition published by Vantage Books, 1976.

Kant, Immanuel. 1790. *Critique of Judgement*, Reprinted in *Great Books* 39: 461. Chicago: Encyclopedia Britannica, 1993.

Keller, Evelyn Fox. 1990. "Physics and the Emergence of Molecular Biology." *J. Hist. Biol.* 23(3): 389–409.

Keller, Evelyn Fox. 1995. *Refiguring Life*, New York: Columbia University Press.

Keller, Evelyn Fox and Jeremy Ahouse. 1997. "Writing and Reading about 'Dolly'." *BioEssays* 19 (8): 741–742.

Kitcher, Philip, and Wesley Salmon, eds. 1989. *Scientific Explanation*. Minneapolis: University of Minnesota Press.

Lodish, Harvey. 1995. Contribution to "Through the Glass Lightly" section. *Science* 267: 1617.

Meyerowitz, Elliott. 1994. "The Genetics of Flower Development." *Scientific American* 271(5): 40–47.

Moss, Lenny. 1992. "A Kernel of Truth? On the Reality of the Genetic Program." *PSA* 1: 335–48.

Spemann, Hans. 1907. "Zum Problem der Correlation in der tierischen Entwicklung." *Verhandl. deutsche zool. Gesell.* 17: 22–49.

Spemann, Hans. 1938. *Embryonic Development and Induction.* New Haven: Yale University Press.

Strohmann, Richard C. 1997. "The Coming Kuhnian Revolution in Biology." *Nature Biotechnology* 15: 194–200.

Tautz, Diethard. 1992. "Redundancies, Development and the Flow of Information." *BioEssays* 14(4): 263–266.

Thomas, James H. 1993. "Thinking about Genetic Redundancy." *Trends in Genetics* 9(11): 395–399.

Waddington, C. H. 1940. *Organizers and Genes.* Cambridge: Cambridge University Press.

Waddington, C. H. 1948. "The Genetic Control of Development." *Symp. Soc. Exp. Biol.* 2: 145–154. New York: Academic Press.

Waddington, C. H. 1957. *Strategy of the Genes.* London: Allen & Unwin.

Waddington, C. H. 1962. *New Patterns in Genetics and Development.* New York: Columbia University Press.

Weinberg, Robert. 1985. "The Molecules of Life." *Scientific American* 253(4): 48–57.

Wilkins, Adam. 1992. *Genetic Analysis of Animal Development.* New York: Wiley-Liss.

Wilkins, Adam. 1997. "Canalization: A Molecular Genetic Perspective." *BioEssays* 19(3): 257–262.

Wolpert, Lewis. 1994. "Do We Understand Development?" *Science* 266: 271–272.

Woodward, James. forthcoming. *A Theory of Explanation.*

Chapter 12

Toward an Epistemology for Biological Pluralism

HELEN E. LONGINO

WHY BE A PLURALIST?[1]

During the heyday of logical empiricism, a radical form of unificationism was in fashion. "The Unity of Science as a Working Hypothesis," by Oppenheim and Putnam (1958), was one of the most bold and radical expositions of the view. It proposed that all (true) scientific theories could be hierarchically ordered such that each rung in the hierarchy would be derivable from the preceding one (plus some definitions or connecting sentences). The most fundamental level was occupied by theories in physics, the most derived was occupied by social sciences. The relation of each level to that from which it was derived was called reduction. Thus sociology was reduced to psychology, psychology to biology, biology to chemistry, and chemistry to physics.

Such a picture has both metaphysical and epistemological (or substantive and methodological) aspects. Metaphysically, all entities and processes are, when fully analyzed, identical with (collections of) the entities and processes that figure in the laws of physics. Epistemologically, all genuine knowledge of the natural world is ultimately expressible in the terms of the physical sciences. Given the formal way in which reduction was understood, this poses significant constraints on the forms of knowledge in the other sciences, as well as on methods of investigation. These constraints include unity within a discipline and the comparability and interconvertibility of observational statements and measurements. It is in many ways a compelling picture, especially if one thinks that there is, after all, just one reality, and that the job of the sciences is to make sense of that one reality. The picture has, however, become harder to maintain. First the formal aspects came into disrepute, with the general disintegration of the positivists' formalist program. And then philosophers of the other special sciences rejected both the pride of place accorded to physics and the assumption that eliminability of a discipline's theories (apart from physics) was a criterion of its scientific character. Although there are philosophers like Philip Kitcher who see unification as a central aim of scientific inquiry (cf. Kitcher 1993), many philosophers of science are now embracing forms of pluralism. Patrick Suppes has argued, in the context of what he calls probabilistic metaphysics, for a pluralistic understanding of scientific knowledge. The multiplication of subdisciplines with their own specialized vocabularies, he argues, spells doom for the radical unification envisioned by Oppenheim and Putnam (Suppes 1984). John Dupré has argued, drawing partly on biology, for the radical disunity of science and an accompanying metaphysics he calls promiscuous realism (Dupré 1993). Nancy Cartwright has argued that not even physics displays the kind of unity presupposed as an ideal (Cartwright 1995). Nevertheless, it is contemporary biology which has stimulated most contemporary pluralist thinking.

The biological sciences have been the sites of fierce debate over both theory and method at least since Darwin. While earlier in this century these debates were quite general – mechanism versus vitalism, for example – current debates tend to be conducted within or about particular subfields. Evolutionary theory is the site of debates over the units and levels of selection. Behavioral biology is not only the site of debates about nature versus nurture, but also encompasses multiple biological approaches – genetic, biochemical and physiological, neurological – which may or may not be reconcilable in an integrated theory. In cell biology, scientists debate the extent of the regulative role of DNA. In ecology, too, researchers debate both conclusions and the methodologies used to reach them. While these disagreements involve both substantive and methodological issues, usually one aspect is the focus of debate in any given instance. Sometimes, potentially conflicting theories or methods either do not give rise to debates or else give rise to debates that do not fully engage the researchers themselves or engage only some members of the relevant fields. This suggests that whether or not different theories and methods are drawn into debate depends on factors other than the difference, or even the incompatibility, between them. Furthermore, the practical conduct of inquiry does not wait upon the resolution of such debates, nor does it require consensus on a single theory or method. This is what Suppes was pointing to in his brief for pluralism. In offering specific examples of difference within subfields, I will follow C. K. Waters in speaking of researchers' *approach* (Waters forthcoming). An approach is something like Kuhn's paradigms, but less encompassing. As I think of it, an approach is characterized by a kind of question and a set of investigative tools used to develop answers to the question(s). Question and tools together often presuppose a theoretical claim, or at the very least a model of the portion of the world being investigated. But the establishment of the theoretical claim is not the main aim of the research approach presupposed. The aim is the answering of the question.[2] The following sketches show some of the situations in which pluralism of approach has been or should be advocated.

Example 1

The behavioral sciences are the most obvious place to look for differences and debates about the best way to study behavior. Even setting aside such highly controversial issues as the validity of psychoanalytic theory or of human sociobiology, contemporary biology hosts multiple approaches to behavior. These include the genetic, the environmentalist, the developmentalist, and the biochemical. The subfield of behavioral genetics is implicated in at least two distinct debates. One of these is a variation on the familiar nature-nurture debate. Another concerns the relative merits of genetic as opposed to developmental approaches to investigating and explaining behavior. In both cases, debate proceeds both with respect to particular behaviors and with respect to overall approach.

Proponents and practitioners of behavioral genetics are asking what proportion of the variation in a trait in a population is owing to variation in genes and what proportion is owing to variation in environment. Their environmentalist opponents ask what social or environmental circumstances nurture a disposition for and elicit the expression of a given trait.[3] In attempting to understand why some individuals are more aggressive than others, the behavior geneticist uses heritability measures and relies on twin and adoption studies to assist in partitioning variance. The environmentalist looks instead for common and differentiating factors in the family and school environments of aggressive youth. The behavioral geneticist argues that studying correlations of factors within families confounds environmental and genetic sources of variation. The environmentalist argues that twins and adoptees are not representative of most families or most individuals and hence are an inappropriate basis for generalization. They also differ as to the differentiation of environments, one arguing that the primary differentiation is between abusive and nonabusive environments, another arguing that finer discriminations are required. Given that the two sides agree that etiology is ultimately located in both genes and environment, it is clear that they could pursue their separate research programs, each investigating behavior modeled as their approach assumes, without engaging in debate. The debate is forced on them by factors other than, or in addition to, their nonconsilience regarding particular issues.

The debate between behavior geneticists and developmental (or a subset of developmental) scientists takes a slightly different form.[4] The developmental scientist is interested not in partitioning variance, but in understanding the proximate mechanisms of phenotypic development. Where the behavioral geneticist and the environmentalist are interested in accounting for differences, the developmentalist is interested in accounting for similarities: not differences in feeding or mating behavior within a population, but commonalities like species-specific patterns of vocalization. In the case of aggression, the behavior genetics question would be how much of the difference between aggressive and nonaggressive individuals is explained by genetic difference and how much by environmental difference.[5] The developmental question would be how individuals identified as aggressive become that way. In spite of the self-ascribed differences in explananda, all three approaches really cast light on both difference and similarity. What makes us similar to the others in a given set is what differentiates us from those outside that set. What really differs is the focus of investigation (or approach). Geneticists end environmentalists alike design studies around differences in a population;

developmentalists may attempt to induce differences in order to understand the mechanisms of development of a trait common to a set of individuals.

Developmental systems thinkers favor an approach that not only integrates but shows the interaction of genetic, physiological, and environmental factors. This approach tends to stress the importance of what is often dismissed as "noise," a nondeterminable result of the coaction of the varied factors, and argues that those factors cannot be partitioned as the geneticist hopes. The geneticist retorts that tracing proximal causal mechanisms in behavioral development is too complicated – that the research questions must be broken down and simplified in order to make progress in understanding behavior. Again, it seems that the research programs of each camp could be pursued in parallel; that debate or competition arises only when either of them claims exclusive or most promising access to knowledge or understanding. But this is not a necessary part of pursuing the questions that each program poses to behavior. Certainly one of the factors contributing to the competition is the perceived social implications of the various approaches. It is not an accident that a significant proportion of this research literature is reviewed in criminology journals. And, in addition to possible practical applications, overall political orientations tend to include the support of one or another of these approaches.

Biochemical or physiological approaches look for chemical correlates of aggression. Currently serotonin levels are a popular topic of research.[6] There is often an assumption that differences in the production or metabolism of such factors as hormones or neurotransmitters have a genetic base, minimizing the debate with geneticists. This assumption is not warranted, however, since these physiological phenomena are also responsive to environmental factors. Some years ago, circulating testosterone was thought to be causally involved in higher levels of aggression. It turned out, however, that the causal relation was the reverse: that circulating testosterone levels in males tended to increase as a consequence of stress, whether the individual was the initiator or recipient of aggression, and that stress was associated with position in dominance hierarchies. Thus, biochemical approaches must be understood as yet a fourth way to investigate behavior whose capacity to produce interesting and informative results does not depend on its being integrated with one of the other three.

Example 2

A more restricted, but no less revealing, example of epistemologically charged difference is offered by Nicholas Rasmussen's work on electron microscopy (Rasmussen 1995). In the 1950s two groups of researchers using electron microscopes to study mitochondria disagreed about the internal structure of these organelles. Mitochondria look like elongated ovals. One group argued that perceived transverse ridges inside the mitochondrial walls were partial, leaving a free channel through the length of the interior. The other group argued that at least some of the transverse structures were plate-like membranes extending across the whole diameter of the mitochondrion, thus blocking any free channel. The groups also disagreed about the relation of these internal structures to the external or surrounding membrane. These disagreements over structure were a function of disagreements over how to interpret and evaluate the electron micrographic evidence. Each group had micrographs that supported its theory of structure. But the micrographs yielded different images of the mitochondria as a result of the use of different fixating and other sample preparation techniques. One group argued that micrographs should be assessed by criteria internal to micrography, such as the degree of resolution of the image. The other group argued that consistency with data from other sources (fractionation biochemistry, in this case) was at least equally, and perhaps more, relevant to determining which of conflicting images was the best representation of mitochondrial structure. This second group succeeded in having its micrographs accepted as the standard, even though they were by micrographic standards inferior.

Example 3

As the discipline of ecology has turned to experimental methods, differences have arisen regarding both the proper kind of experimentation and the merits of experiment at all, in contrast to field observation.[7] One recent debate concerned worldwide drops in amphibian populations. A research group in Oregon conducted an experiment with paired sets of frog eggs, one set exposed to and the other set protected from ultraviolet radiation. The protected eggs developed into tadpoles at a much greater rate than did the exposed ones, from which the researchers concluded that increased UV radiation was the culprit in the observed decline of amphibians. Dissenters claimed that there had been no investigation into whether UV radiation had increased over the time amphibian mortality was on the rise, and that only a long-term observation of an ecosystem and its inhabitants could ascertain the cause of mortality in the field as opposed to in the laboratory. Even among experimentally oriented ecologists, there are differences as to the relative value of ecosystem experimentation (e.g., seeding patches of ocean with iron) and experimentation with model systems like the paired sets of frog eggs in the laboratory, or even more complicated but still constructed and controlled systems. Ecosystem experimentation which manipulates variables in actual environments is more realistic, but model systems, while both deliberately and unintentionally simplified, offer better opportunities for identifying and controlling possible causal factors. Given that different (and different levels of) elements are present in the two kinds of experimental regime – even when one is intended to be a model of the other – it is likely that different studies of the effects of introducing the same agent will yield different quantitative results. Advocates of the ecosystem approach applaud its value for evaluating and predicting the effects of environmental change. Advocates of the model approach stress the replicability and control of relevant parameters that the approach offers.

Example 4

Philosopher of biology Sandra Mitchell has examined the variety of explanations of the division of labor in social insect colonies. There are both evolutionary and ontogenic explanations, and among the ontogenic, or self-organization, explanations there are a variety of causal models of the division of labor. One appeals to genetic diversity among the individual colony members, which results in different response thresholds to a given stimulus, which results in turn in a division of labor. Another proposes that all colony members are born with an identical work algorithm, but that the nest architecture varies the stimulus received, thus evoking different behaviors from different individuals. While some would see this as a matter of incompatible explanations requiring resolution through the choice of one, Mitchell argues that these models do not need to be seen as incompatible in that sense. They are idealizations, and as such map onto an ideal world rather than the real world. In applying them in the actual world, we might say that each applies but in different circumstances. Some species accomplish the division of labor in one way and others in a different way. Thus, there cannot be a single model of division of labor in social insects. Nature produces the same ends with different means. Natural complexity, according to Mitchell, stands in the way of a unified science and requires pluralism.

Example 5

Ken Waters implicitly endorses a stronger form of pluralism in his analysis of the units of selection controversy. Biologists and philosophers of biology have argued over whether selection acts on genes, genotypes, individual organisms, or groups of organisms. Waters claims that the parties to these debates who are also realists have made two unexamined assumptions:

1. for any selection process there is a uniquely correct description of the operative selection forces and the level at which each impinges; and
2. population level causes must enhance the probability of their effects in at least one context and not decrease it in any other.

The first assumption fuels debates among theorists committed to analysis of selection at different levels. The second is used by opponents of genic selectionism to discredit that view. Waters analyzes cases and arguments offered by Williams, Dawkins, and Sober and Lewontin, with the aim of showing that different selection theorists parse the domain and its environment in different ways. This parsing, for which no claims of unique correctness can be claimed, supports the identification of different levels as the one at which selec-

268

tion forces are acting. A similar analysis emphasizing the heterogeneity of real environments supports Waters's contention that *all* selection forces are context-sensitive. He advocates instead what he calls tempered realism, which transforms the above two assumptions as follows:

1A (TR) Accounts which differ as to the level on which selection is operating can nevertheless accurately represent the same causal network in different ways.
2A (TR) Population level forces can increase the chances of their effects in some contexts and decrease them in others.

1A (TR) effectively says that the selective forces and units selected can be differently but equally correctly represented by different theories of the same process. And the principle used to knock out one level of selection has to be abandoned, and genic selection theories tolerated, when the parsing of domain and environment requires it. Tempered realism is pluralism in all but name.

Thomas Kuhn, in his discussion of the succession of theories, claimed that only with the emergence of a single hegemonic paradigm could a field of inquiry mature into a science. If we apply that view to the examples above, we have to say that we are dealing with nonsciences or immature sciences. But one has to ask on what kinds of grounds this can be asserted. Kuhn, of course, was talking primarily about physics, and perhaps about the history of physics represented as a certain kind of undertaking: the development of an encompassing theory of, say, motion. Philosophers of biology who advocate pluralism, especially those advocating strong forms of pluralism, are claiming that, whatever may be the case in physics, the complexity of organic and living entities and processes eludes complete representation by any single theoretical or investigative approach. Any given approach will be partial, and completeness will be approached not by a single integrated theory, but by a plurality of approaches that are partially overlapping, partially autonomous, and resisting reconciliation. Given that a phenomenon is modeled slightly differently in different approaches, quantitative measures will vary between approaches, so that comparing data descriptions will show inconsistencies between the approaches. Deep insight into a phenomenon will be purchased at the cost of losing unifiability with other approaches.

The difficulty with such pluralism, of course, is that it sits uneasily with standard ways of thinking about knowledge or about the world. If

1. different approaches or investigative strategies presupposing different theories or models result in non-reconcilable measures of what pretheoretically would be identified as the same phenomenon, and
2. there is one world and not many, and
3. all true statements must be consistent with each other, then

those theories cannot all be true, and some investigative strategies are simply wrongheaded. But by what criteria are we to evaluate investigative strategies? And if incompatible strategies nevertheless produce usable results, why should we insist upon the relinquishing of one or the other? Not only is the objectivity of the sciences at stake, so seemingly is the metaphysical unity of the world. How can we make sense of multiple, equally good theories (or approaches) without slipping into multiple-world talk? Even Kuhn ended up suggesting that scientists holding sufficiently incommensurable yet incompatible theories, occupied different worlds. If meant literally, this is a recipe not for the disunity of science but for the fragmentation of sciences and worlds. The pluralism envisioned by theoretical pluralists is a pluralism of theories of a singular world.

EPISTEMOLOGY, PLURALISM, AND SCIENTIFIC INQUIRY

Can our conceptions of knowledge and inquiry accommodate such a pluralism? I would like to suggest several guidelines to observe in thinking about this question.

1. The plurality of representations in biology may be a function of how the world is, or of human intellectual equipment for and interests in understanding the world. Our epistemology cannot dictate which, thus

2. A satisfactory epistemology should be open to theoretical plurality *or* theoretical unity being the final result of inquiry (as
   though there were any such thing!).
3. The issue of theoretical pluralism ought not be decided by one's
   choice of epistemology.
4. A suitable humility requires a modest epistemology. An epistemology – as a theory of human knowledge – ought not to promise
   complete knowledge (or trade in other absolutes, like certainty)
   but ought to give sense to the distinctions and normative judgments that are a part of epistemic discourse.

It is tempting to think that scientific knowledge is like ordinary
knowledge except better. But scientists are not (or not just) better
observers and more careful reasoners than the rest of us, they do
something different. The purpose of scientific inquiry is not only to
describe and catalog, nor even explain, that which is present to
everyday experience, but to facilitate prediction, intervention, control, and other forms of action on and among the objects in nature.
Description and classification are in service to these more overarching purposes, which move the focus of inquiry away from what is
present to us to the principles, processes, and mechanisms that produce or underlie what is present to us. This involves making visible
(e.g., via dissection or vision-enhancing instruments) what is invisible; in other cases it involves postulating processes and mechanisms
beyond the range of human sensory capacities – too long or too short
in duration, too big or too small to be perceived, outside of the
frequency ranges to which human senses are open. In most cases
scientists conduct these activities in and on idealized situations,
whether in the laboratory or in thought experiments, so as to be able
carefully to examine one aspect of a process rather than be stymied
by attempting to render its full complexity. This way of experiment is
arguably what most distinguishes modern Western science from
other modes of knowledge and inquiry.

Modern Western science is also characterized, among other things,
by the postulation not just of underlying processes and mechanisms,
but of processes and mechanisms that facilitate human intervention in
the natural world. This is manifest in the kind of science first fully
developed in the modern period – mechanics – and in the close ties

that have developed between Western productive capacities and scientific knowledge. In spite of this concern with what underlies experienced processes, the establishing of theories, in the sense of large-scale systematic accounts of a range of related phenomena, is not at the center of most scientists' work. Rather they presuppose theories, and defend them when challenged, but as a distraction from their central mission, which is to answer (in light of general theoretical frameworks) specific, carefully articulated questions addressed to the idealized situation of the physical or mental laboratory or to some carefully delimited portion of the natural world.

The gap between what is present to us, whether in the kitchen and drawing room or in the laboratory, and the processes that we suppose produce the world as we experience it – between our data and the theories, models, and hypotheses developed to explain the data – has been at the heart of philosophical reflection about scientific knowledge. Once the logical empiricists abandoned the notion that all meaningful theoretical statements could be fully translated into observational statements, the problem of characterizing the relation between observation and theory haunted philosophy of science. Thomas Kuhn (1962), Paul Feyerabend (1962), and Russell Hanson (1958) emphasized the ways in which theory influenced observation. They introduced the notion of the theory-ladenness of observation, that is, that observation is not neutral, but is permeated by theoretical commitments. The meaning of scientific terms, too, was held to be theory-laden. This closes the gap between observation and theory, but at the price, as many philosophers protested, of circularity. Philosophical discussion of this notion seems to have stabilized in agreement that observation reports both depend on and imply theoretical commitments, but that as long as the theory that loads the observation or description of data is not the theory for which the data are invoked as evidence, circularity is avoided (Hesse 1980: 63–110).

As long as the content of theoretical statements is not represented as generalizations of data and the content of observational statements is not identified with theoretical claims, however, then the gap between hypotheses and data persists, and the choice of hypothesis is not fully determined by the data. Nor do hypotheses specify the data that will confirm them. Data alone are consistent with different and conflicting hypotheses and require supplementation. This problem is distinct from the traditional problem of justifying induction, which concerns the relation between a generalization and its instances. Pierre Duhem, the first philosopher of science to raise the underdetermination problem, emphasized assumptions about instruments – for example, that a microscope has a given power of resolution, or that a telescope is transmitting light from the heavens and not producing images internally, or not systematically distorting the light it receives. But the content of background assumptions also includes substantive (empirical or metaphysical) claims that link the events observed as data with postulated processes and structures. For example, that two kinds of event are systematically correlated is evidence that they have a common cause or that one causes the other in light of some highly general, even metaphysical, assumptions about causality. The correlation of one particular kind of event – for example, exposure to or secretion of a particular hormone – with another – a physiological or behavioral event – is evidence that the hormone causes the physiological or behavioral phenomenon in light of an assumption that hormones have a causal or regulative status in the processes in which they are found, rather than being epiphenomenal to or effects of those processes. Such an assumption has both empirical and metaphysical dimensions. Assumptions of this kind establish the evidential relevance of data to hypotheses. In the language being used here, they provide a model of the domain being investigated that permits particular investigations to proceed.

Background assumptions, then, include substantive and methodological hypotheses which, from one point of view, form the framework within which inquiry is pursued and, from another, structure the domain within which inquiry is pursued. These hypotheses are most often not articulated, but presumed by the scientists relying on them. They facilitate the reasoning between what is known and what is hypothesized. From a traditional perspective this raises major problems for justification: if the justification of hypotheses requires assumptions, then how are these assumptions in turn justified? And how is it possible to screen out biasing factors such as individual idiosyncrasies, wishful thinking, values, social prejudices, ideologies, metaphysics? In some cases evidence for assumptions can be offered, but the same problem of underdetermination besets such evidential reasoning as well.

Underdetermination shows that any empirical reasoning takes place against a background of assumptions which are neither self-evident nor logically true.[8] The identification of good reasons is similarly context-dependent, whether this is accomplished by the scientist in action or by the philosopher in reflection. Some sociologists of science have used versions of the underdetermination problem to argue that epistemological concerns with truth and good reasons are irrelevant to the understanding of scientific inquiry (Barnes, Bloor, and Henry 1996; Pickering 1984, 1985; Shapin 1994; Collins and Pinch 1993; Knorr-Cetina 1992 and forthcoming; Latour 1987, 1993). Rather than spelling doom for the epistemological concerns of the philosopher, however, the logical problem of underdetermination, together with the studies of laboratory and research practices, changes the ground on which philosophical concerns operate. That is, the philosophical concern with justification is not irrelevant, but must be reconfigured somewhat in order to be made relevant to scientific inquiry. One compelling reconfiguration involves treating justification not just as a matter of relations between sentences, statements, or the beliefs and perceptions of an individual, but as a matter of relations within and between communities of inquirers.

If the underdetermination problem makes evident the role of assumptions in evidential relations, an account of justification must have something to say about them. Simply requiring more evidence produces an endless regress of reasons. The socializing move instead introduces criticism and the survival of criticism as key aspects of justification. This locates justification not just in the testing of hypotheses against data, but in subjecting hypotheses, data, reasoning, and background assumptions to criticism from a variety of perspectives.[9] Establishing what the data are, what counts as acceptable reasoning, which assumptions are legitimate and which not, becomes a matter of social interactions as much as a matter of interaction with the material world. Since assumptions are, by their nature, usually not explicit, but taken-for-granted ways of thinking, the function of critical interaction is to make them visible as well as to examine their metaphysical, empirical, and normative implications.

This socializing of justification has two consequences. In the first place, any normative rules or conditions must include social interactions. A full account of justification or objectivity must spell out conditions that a community must meet for its discursive interactions to constitute effective criticism. I have elsewhere proposed that establishing or designating appropriate venues for criticism, uptake of criticism (i.e., response and change), public standards that regulate discursive interactions, and equality of intellectual authority are conditions that make transformative criticism possible (Longino 1990: 62–82). As conditions that regulate the role of subjectivity in inquiry, they are also conditions of objectivity. Public standards can include aims and goals of research, background assumptions, methodological stipulations, ethical guidelines, and so on, and are themselves subject to critical scrutiny. Secondly, even though a community may operate with effective structures that block the spread of idiosyncratic assumptions, those assumptions that are shared by all members of a community will not only be shielded from criticism, but, because they persist in the face of effective structures, may even be reinforced. One obvious solution is to require interaction across communities, or at least to require openness to criticism both from within and from outside the community. Here, of course, availability is a strong constraint. Other communities that might be able to demonstrate the non-self-evidence of shared assumptions or to provide new critical perspectives may be too distant – spatially or temporally – for contact.[10] Background assumptions then are only provisionally legitimated; no matter how thorough their scrutiny given the critical resources available at any given time, it is possible that scrutiny at a later time will prompt reassessment and rejection. Such reassessment may be the consequence not only of interaction with new communities but also of changes in values or other assumptions within a community.

Using this social account of justification one might then say:

A theory T or hypothesis H is epistemically acceptable in community C at time *t* if T or H is supported by data *d* evident to C at *t* in light of reasoning and background assumptions which have survived critical scrutiny from as many perspectives as are available to C at *t*, and the discursive structures of C satisfy the conditions for effective criticism.

Clearly, then, that a theory is acceptable in C at *t* does not imply that C will not come to abandon it at some later time. Furthermore, there is no requirement that members of C reject background assumptions simply because they are shown to be contingent or to lack firm support. Unless background assumptions are shown to be in conflict with values, goals, or other assumptions of C, there is no obligation to abandon them – only to acknowledge their contingency and thus to withdraw excessive confidence. Background assumptions are, along with values and aims of inquiry, the public standards that regulate the discursive and material interactions of a community. The point here is that they are both provisional and subordinated to the overall goal of inquiry for a community. Truth *simpliciter* cannot be such a goal, since it is not sufficient to direct inquiry. Rather, communities seek particular kinds of truths (representations, explanations, technological recipes; truths about the development of individual organisms, about the history of lineages, about the physiological functioning of organisms, about the mechanics of parts of organisms, etc.) which are determined by the kinds of questions they are asking and the purposes for which they ask them, that is, the uses to which the answers will be put.

There seem to be at least two choices one can make about the use of the term 'knowledge' in a dynamic and social epistemology of this sort. If one wants to count a community's acceptable theories as knowledge, one can treat knowledge as provisional, partial, and context dependent. But those uneasy with the instability this permits may propose to disqualify the provisional and the context-dependent as objects of knowledge. This might eliminate theories and models as objects of knowledge, leaving as knowable only the observational data which support them. This criterion is more draconian than it seems. While the observed or measured character of data warrants a different level of doxical confidence, there are, however, still important senses in which they, too, are provisional and partial. Different theoretical frameworks can make different data, different descriptions of data, different aspects or measurements of the data, salient. Different statements about observation will be meaningful and relevant in different theoretical contexts. Some descriptions of celestial bodies, while salient to sixteenth-century astronomers and twentieth-century astrologers, are of no interest to contemporary astronomers. The measurements of contemporary astronomers may prove to be as useless to astronomers of the twenty-fifth century as the earlier measurements are to us. Systems of measurement change, criteria of good measurement change, the kinds of relationships among data that are important change, the kinds of data that are important change. So if, as on the first alternative, we deny that the provisional counts as knowledge, then little or nothing remains as knowledge. Since we cannot predict which of our beliefs will persist and which wither into disuse, almost nothing we count as knowledge can be ascribed permanence or certainty. The temptations to legislate provisionality away and the untoward consequences of doing so suggest that our concept 'knowledge' contains elements that are brought into tension with each other when confronted with science and its history. This tension necessitates decisions of a semantic nature that will utilize some aspects of the concept at the expense of others.

Whatever semantic decisions are made about knowledge, it is clear that epistemological analysis of scientific theory and inquiry must include analysis of the social and intellectual context in which inquiry is pursued and theories and hypotheses are evaluated. The intellectual context is constituted of background assumptions and investigative resources – instruments, samples, experimental protocols, and so forth. The social context is the set of institutions and interactions in and through which assumptions and resources circulate, as well as the larger social environment in which institutions and interactions are embedded. The analysis of objectivity referred to above states the (kinds of) conditions such institutions and interactions should meet. Meeting these conditions, however, does not permit detaching justification or acceptability from the context within which justification is achieved. Thus, attention to the intellectual context is important for appreciating the scope and limits of any given claim.

The assumptions which partially constitute this context are of at least two kinds: substantive and methodological. Substantive assumptions concern the way the world one is investigating is. They may be compositional or processual. An example of the former is the assumption that the material world is constituted of particles which at the most fundamental level are indivisible. A processual assumption that had a following recently is the assumption that all biological development is controlled by genes. Methodological assumptions have to do with the means we have of developing and acquiring knowledge. They can range from general philosophical views, like the commitment to some form of empiricism, to quite particular views about the kinds of data appropriate for kinds of question: field observation versus experiment, animal models versus human studies, in vivo versus in vitro studies, and so on. They include assumptions concerning how much data of a certain type should be required, what kinds of data are relevant, and what mix of kinds of data from different kinds of study techniques should be required. They also include what philosophers sometimes call epistemic or cognitive values. The most commonly cited of these are simplicity, consistency, explanatory power, and accuracy, but their contraries – heterogeneity, complexity, and social utility – also have their advocates. The range of methodologies obviously depends in part on the availability of investigative resources, but these are not sufficient to limit the number of methodologies to one.

Given such an array, one can understand investigative, or scientific, communities as constituted around selections of substantive and methodological assumptions. These selections are a function of both the aims of research and inherited tradition. To call them selections does not mean that they are deliberately picked out from a possible assortment, but that they represent a subset of possible alternatives, reliance on which can be defended either by reference to the goals of the particular research community or because "that's the way we do things." This latter form of justification, the appeal to tradition, is more fragile than the former. All of these – substantive assumptions, methodological assumptions, aims and goals, and the arguments linking assumptions and practices to aims and goals – are subject to critical scrutiny, debate and defense, and all of them may bear complicated relations to the social, political, and aesthetic values of a community's cultures.

That set of methodological choices, commitments, or standards can be called a community's epistemology. Characterization of such an epistemology would include a specification of a number of elements in addition to specific methods relied on to produce answers to questions and standards of necessity and sufficiency of various kinds of data. A full characterization would include a specification of the kind of knowledge sought: causal or processual (e.g., how x works), compositional (e.g., what y is made of), distributive (e.g., what is the frequency of occurrence of *z*), or descriptive (e.g., the number of species in an ecosystem, the boiling point of *s*). Causal questions may, furthermore, be either mechanical or historical. The specification of kind of knowledge sought might also include formal characteristics, for example, how general the knowledge is that is sought. This much of an epistemology is what was referred to earlier as an approach: the application of an epistemology to a particular domain. A complete characterization of a local epistemology would also provide some account of why the methods used are thought adequate in relation to the kind of knowledge sought. Such an account will typically involve reference to substantive assumptions and cognitive values, but there may also be pragmatic and/or professional reasons to choose particular methods or standards of evidence. Such reasons may include the availability of necessary instruments, or the unavailability of alternative procedures; they may involve issues of communicability and standardization, acceptance within related communities, and so on.

If methodological rules and procedures are not claimed to be self-evident and context-independent, then their rationale, as already stated, must lie in the aims and goals of the inquiring community (or in its traditions). Identifying these features of a local epistemology, particularly the assumptions and values that link methods to kinds of knowledge sought, is a matter not just of picking out the methods and standards that link data to hypotheses in research articles, but of reconstructing them from an analysis of the context of inquiry: correspondence, accounts of controversy and of interventions in controversy, study of institutional settings, priorities, and constraints. Since individuals may have a variety of motives for engaging in community activities, the values and priorities that organize an approach cannot be identified simply from individuals' values and interests, but must be shown to be both commonly or institutionally endorsed and effective. A local epistemology is a dynamic complex of beliefs, norms, goals, and practices. These elements can change as a community encounters external and internal challenges. External challenges may include interaction with other communities; internal challenges are generated when social goals and values change or when the application of particular norms and practices in concrete situations yields results that are in tension with other elements. Scientific inquiry not only produces answers to questions, but those answers themselves open up new issues for investigation, new landscapes of ignorance. This dynamic aspect of inquiry means that its epistemology, too, must be dynamic.

The localizing of such status as justification and acceptability advocated here is certainly compatible with pluralism, but may leave one puzzling about what epistemology might then consist in. One of the puzzles, of course, is what happens to the prescriptive or normative aspect of epistemology if justification is context-dependent. One solution to this puzzle is to distinguish between local and general epistemology. General or philosophical epistemology becomes an interpretive inquiry, an exploration of the meaning of concepts like knowledge, truth, justification. Normative epistemology is then the province of the local epistemologies I have been describing, whose authoritative reach extends only to those who intend, through membership in a community, to be bound by its strictures. The relation of local to general epistemology is more complex than this division indicates. For example, the status of the conditions of objectivity or transformative criticism is problematic. Whose conditions are these? and what is the ground of their normative claim? If we say that they are a community's conditions, then it is an empirical question whether they form part of a community's standards, or of a community's epistemology, and they have no general claim. But if they are general, then their proposal as normative contravenes the view that normative epistemology is local.[11] I think it is best to treat them as an explication of "objectivity" and thus as the result of an interpretive exercise in general or philosophical epistemology. As norms they can be understood as hypothetical imperatives: "if you wish to be objective, then do X."

THE CASES REVISITED

What does the way of thinking about knowledge and inquiry mean for the examples cited earlier in this chapter? In the case of multiple approaches to the study of behavior, each must be understood in terms of the kind of knowledge sought and the adequacy of methods used to answer the questions characterizing the different approaches. As it turns out, participants in this debate have different conceptions of what counts as scientific knowledge and of the aims of inquiry, and different views about the proper relation of scientific inquiry to social issues. These different normative views ground their different methodologies.

In the ecological case, one can understand the different experimental approaches as driven by differently oriented questions – how elements in a natural system together respond to a particular kind of change, or what effects a given change can produce on a restricted set of elements. Further study of the contexts of inquiry is required to understand how proponents of these different experimental strategies would relate them to the goals of research, the discipline of ecology, the current social and political concerns about the environment, and the multiple institutions within or under the sponsorship of which their research is conducted. Rasmussen's electron microscopy example is interesting because it demonstrates the displacement of a conventional value – greater detail, and hence accuracy – by another – transmissibility and standardization of results across different disciplines. Here one might ask under what circumstances these different values take precedence and what the circumstances were in this case which resulted in assigning priority to standardization. Rather than assume that one party was right and one wrong, the historian or philosopher may be better served by instead studying the operation of the different epistemologies that underlay this conflict about the proper representation of mitochondria. Since standardization means the stabilization of a representation and a consequent broadening of our ability to act on the object represented, pressures for action on environmental problems may produce similar dynamics in the ecological context.

Mitchell argues that nature's diversity and complexity mean that there are multiple accounts of the division of labor in social insect colonies. She does think, however, that integration is ultimately possible: each causal story is a model which must be applied to a particular species or subspecies. For each species there is a model that uniquely fits. Waters, by contrast, argues that the same phenomenon can be approached in different ways that yield different representations and explanations, different tracings of causal pathways. Depending on certain starting points, any of several accounts is correct.

These examples indicate that plurality of accounts or representations can arise in multiple ways: the variety of causal pathways to a similar end; focus and emphasis on one aspect of a complex causal process, whether as a function of explicit interest (what is the genetic contribution to x?) or as a consequence of initial parsings of the domain under investigation; use of observational methodologies that of necessity show only that aspect of a process that they are designed to show, or use of evaluation criteria that in a given context are satisfied by inconsistent representations. Finding a uniquely correct account would require identifying the uniquely correct starting point in each of these reservoirs of plurality. But this cannot be an epistemological project without arbitrarily closing off avenues of investigation, thus violating the canons of modesty articulated above. The philosopher interested in understanding the structures of reasoning and justification must examine individual cases and their contextual features which both fill out those structures and have the effect of localizing the epistemological judgments.

CONCLUSION

Waters argues that realism can only be tempered realism, which admits multiple, different, but equally correct accounts of the real. Tempered realism, or realism and pluralism together, also require epistemological pluralism, that is, a philosophical epistemology that recognizes the local character of prescriptive epistemologies associated with particular approaches. Not every way of assembling the elements of such local epistemologies will be successful. The test, however, is not conformity to some higher-level set of rules or principles, but the ability of an epistemology to help a community achieve the understanding it seeks.

The pluralist must also say, I think, that it makes no sense to detach measurements and data descriptions from the contexts in which they are generated, or that as soon as one does, one creates a new context relative to which they are to be assessed and understood. On the view I have set out here, there is no a priori requirement that different approaches to studying the same general area must yield compatible observation statements or measurements or else only one of those different approaches can be correct. The only general requirements would be hypothetical norms emerging from the contextualist analysis of evidence and evidential relations, such as "if you wish a claim to be rationally justified, then the claim and/ or the approach within which it is advanced must be able to meet criticism." Meeting criticism can involve showing how the claim is vindicated by aspects of the chosen approach, showing how the approach satisfies certain goals and then perhaps defending these goals, or modifying the claim or aspects of the approach – for example, by moderating or qualifying its scope or altering aspects of the research process itself.

This leaves unanswered whether critical interaction could lead in the long run to ultimate integration within fields and among fields. But if there is no complete view, then communities (and members of communities) must be free to pursue their own cognitive and social interests in investigative approaches. If any community (whether local or an agglomeration of local communities) is to achieve the best understanding possible for it, then pluralism, not monism, is required. And if there is a single complete view to be had, its achievement will be the end of inquiry and epistemology alike, whether one is a pluralist or a monist.

NOTES

1. I am grateful to Garland Allen, John Beatty, and Jane Maienschein for inviting me to think about these issues as a participant in the Dibner Summer Seminar in the History of Biology in Summer 1996; to Ken Waters for ongoing discussion of them; and to Richard Creath and Ken Waters for their helpful comments on an earlier draft of this chapter.
2. This presupposition of a model encompasses what in Longino (1990) I described as the presupposition of an object of inquiry or of an explanatory model.
3. For a good example of disputation between behavior geneticist and environmentalist approaches see the exchange between Sandra Scarr and Diana Baumrind (Scarr 1992, 1993; Baumrind 1993). See also Plomin, Owen and McGuffin (1994) and Scarr (1987).
4. For a good example of the debate here, see the exchange between Gilbert Gottlieb and his critics (Gottlieb 1995; Burgess and Molenaar 1995; Scarr 1995; Turkheimer, Goldsmith, and Gottesman 1995).
5. While advocates of genetic and environmental approaches are often drawn into debate, bringing other approaches such as developmental systems into the comparison shows how similar they are. They agree

that genes and environment interact in the causation of behavior. They debate over which has the greater causal role.

6. See, for example, the work of E. F. Coccaro et al. (1990, 1992) and of R. J. Kavoussi et al. (1994).

7. For a general account, see Roush (1995). For advocacy of ecosystem experimentation, see Carpenter et al. (1995); for model systems, see Lawton (1995).

8. For a reaffirmation of the problem of underdetermination against a variety of recent attempts to defuse it, see Potter (1996).

9. This expansion or modification of the concept of justification in science is part of a more general move that sees justification and the having of reasons as analyzable through the rubric of response to challenge rather than a static logical structure. See Annis (1978) and Cohen (1987).

10. Richard Creath asks whether this might imply a role for history of science and of philosophy. Certainly, contact among different cultures and traditions has facilitated mutual critical interaction in this century; there is no reason specialists in reconstructing the thought of another time should not generate further bases for critical interactions.

11. I am grateful to Robert Audi for presenting me with this dilemma.

REFERENCES

Annis, David. 1978. "A Contextualist Theory of Justification." *American Philosophical Quarterly* 15: 213–219.

Barnes, Barry, David Bloor, and John Henry. 1996. *Scientific Knowledge: A Sociological Analysis.* Chicago: University of Chicago Press.

Baumrind, D. 1993. "The Average Expectable Environment Is Not Good Enough: A Response to Scarr." *Child Development* 64: 1299–1317.

Burgess, Robert and Peter E. M. Molenaar. 1995. "Commentary." *Human Development* 38: 159–164.

Carpenter, Stephen R., Sallie W. Chisholm, Charles J. Krebs, David W. Schindler, and Richard F. Wright. 1995. "Ecosystem Experiments." *Science* 269: 324–327.

Cartwright, Nancy. 1995. "The Metaphysics of the Disunified World." In David Hull, M. Forbes, and R. M. Burian, eds., *PSA 1994.* East Lansing, MI: Philosophy of Science Association.

Coccaro, E. F., G. Steven, and L. Sever. 1990. "Buspirone Challenge: Preliminary Evidence for a Role for Central 5-HT_{1a} Receptor Function in Impulsive Aggressive Behavior in Humans." *Psychopharmacology Bulletin* 26: 393–405.

Coccaro, E. F., R. J. Kavoussi, and J. C. Lester. 1992. "Self- and Other-directed Human Aggression: The Role of the Central Serotonergic System." *International Clinical Psychopharmacology* 6 (suppl. 6): 70–83.

Cohen, Stuart. 1987. "Knowledge, Context, and Social Standards." *Synthese* 73: 3–26.

Collins, Harry and Trevor Pinch. 1992. *The Golem: What Everybody Should Know about Science.* Cambridge: Cambridge University Press.

Dupré, John. 1993. *The Disorder of Things.* Cambridge, MA: Harvard University Press.

Feyerabend, Paul. 1962. "Explanation, Reduction and Empiricism." In Herbert Feigl and Grover Maxwell, eds., *Minnesota Studies in the Philosophy of Science,* vol. 3 Minneapolis, MN: University of Minnesota Press.

Gottlieb, Gilbert. 1995. "Some Conceptual Deficiencies in 'Developmental' Behavior Genetics." *Human Development* 38: 131–141.

Hanson, N. Russell. 1958. *Patterns of Discovery.* Cambridge: Cambridge University Press.

Hesse, Mary. 1980. *Revolutions and Reconstructions in the Philosophy of Science.* Bloomington, IN: Indiana University Press.

Kavoussi, R. J., J. Liu, and E. F. Coccaro. 1994. "An Open Trial in Personality Disordered Patients with Impulsive Aggression." *Journal of Clinical Psychiatry* 55: 137–141.

Kitcher, Philip. 1993. *The Advancement of Science.* New York: Oxford University Press.

Knorr-Cetina, Karin. 1992. "The Couch, the Cathedral and the Laboratory." In Andrew Pickering, ed., *Science as Practice and Culture.* Chicago: University of Chicago Press.

Knorr-Cetina, Karin. 1998. *Epistemic Cultures.* Cambridge, MA: Harvard University Press.

Kuhn, Thomas. 1962. *The Structure of Scientific Revolutions.* Chicago: University of Chicago Press.

Latour, Bruno. 1987. *Science in Action.* Cambridge, MA: Harvard University Press.

Latour, Bruno. 1993. "One More Turn After the Social Turn." In Ernan McMullen, ed., *Social Dimensions of Scientific Knowledge.* South Bend, IN: University of Notre Dame Press.

Lawton, John. 1995. "Ecological Experiments with Model Systems." *Science* 269: 328–331.

Longino, Helen E. 1990. *Science as Social Knowledge.* Princeton, NJ: Princeton University Press.

Longino, Helen E. 1996. "Cognitive and Non-cognitive Values in Science: Rethinking the Dichotomy." In Lynn H. Nelson and Jack Nelson, eds., *Feminism, Science, and the Philosophy of Science.* Boston: Kluwer.

Mitchell, Sandra. 1995. "Biological Complexity and Theoretical Pluralism." Lecture delivered to the Minnesota Center for Philosophy of Science, University of Minnesota, April 21, 1995.

Oppenheim, Paul and Hilary Putnam. 1958. "The Unity of Science as a Working Hypothesis." In Herbert Feigl, ed., *Minnesota Studies in the*

Philosophy of Science, vol. 2. Minneapolis, MN: University of Minnesota Press.

Pickering, Andrew. 1984. *Constructing Quarks*. Chicago: University of Chicago Press.

Pickering, Andrew. 1995. *The Mangle of Practice*. Chicago: University of Chicago Press.

Plomin, R., M. J. Owen, and P. McGuffin. 1994. "The Genetic Base of Complex Human Behaviors." *Science* 264: 1733–1739

Potter, Elizabeth. 1996. "Underdetermination Undeterred." In Lynn H. Nelson and Jack Nelson, eds., *Feminism, Science, and the Philosophy of Science*. Boston: Kluwer.

Rasmussen, Nicholas. 1995. "Mitochondrial Structure and the Practice of Cell Biology in the 1950s." *Journal of the History of Biology* 28: 381–429.

Roush, Wade. 1995. "When Rigor Meets Reality." *Science* 269: 313–315.

Scarr, Sandra. 1987. "Three Cheers for Behavior Genetics: Winning the War and Losing Our Identity." *Behavior Genetics* 17: 219–228.

Scarr, Sandra. 1992. "Developmental Theories for the 1990s." *Child Development* 63: 1–19.

Scarr, Sandra. 1993. "Biological and Cultural Diversity: The Legacy of Darwin for Development." *Child Development* 64: 1333–1353.

Scarr, Sandra. 1995. "Commentary." *Human Development* 38: 154–158.

Shapin, Steven. 1994. *A Social History of Truth*. Chicago: University of Chicago Press.

Suppes, Patrick. 1984. *Probabilistic Metaphysics*. Oxford: Blackwell.

Turkheimer, Eric, H. H. Goldsmith, and I. I. Gottesman. 1995. "Commentary." *Human Development* 38: 142–153.

Waters, C. Kenneth. 1991. "Tempered Realism about the Force of Selection." *Philosophy of Science* 58: 533–73.

Waters, C. Kenneth. forthcoming. "What Was the Body of Knowledge Called Classical Genetics?" Minnesota Center for Philosophy of Science, University of Minnesota.

Afterword

Biology and Epistemology: Emerging Themes

KENNETH F. SCHAFFNER

INTRODUCTION

The preceding contributions by many of the leading scholars in the history and philosophy of biology are directed at the interactions between the biological sciences and epistemology. As such, they begin to sketch in a neglected area in biological science studies, and also point the way to what are likely to be themes that will occupy scholars during the first decade of the new millennium. In this chapter, I will not attempt to critique and comment in detail on these rich and variform contributions, but rather will identify some of the exciting themes that appear to me to raise profound questions about the nature of the biological sciences and how we come to have and justify knowledge in this area.

As the editors of this volume note in their introduction, the articles fall into roughly three groups that have been placed in their respective sections on (1) the Darwinian theory of evolution and nineteenth-century philosophy of science, (2) historical studies of speciation, development, and neuroscience that examine epistemological disputes, and (3) issues addressing both discovery and justification in more current areas of biochemistry, developmental biology, and Mendelian and population genetics. I will not offer much in the way of commentary on the historical substance of the first group of four essays, since these represent the history of evolution and historical philosophy of science that, though important, falls largely outside this author's expertise. I will, however, touch on several philosophical issues raised in the essays of Hull, Ruse, Hodge, and Richards that resonate with more contemporary debates. I have arranged my reflection under two rubrics in the following sections. I begin with a discussion of empiricism and the New Experimentalism, and then move on to the issues of pluralism, discovery, and explanation in biology.

FROM THE OLD EMPIRICISM TO THE NEW EXPERIMENTALISM

The question whether knowledge can be based on experience or whether it requires some a priori element is thousands of years old. The empiricist-rationalist post-Cartesian "modern" debate, perhaps best typified by the Humean versus Kantian approaches, is a major chapter in that controversy, but the discussion has continued. In the nineteenth century, particularly in relation to the Darwinian theory of evolution, the different philosophical views of Lyell, Herschel, Mill, and Whewell become the main actors in this contentious play. The issue of independent empirical verifiability of causal agents (in Latin, *verae causae*) looms large in the essays of Hull, Ruse, and Hodge in this volume, where they focus on the question whether Darwinian natural selection is a *vera causa*. Perhaps not surprisingly, the empirical weakness of Darwin's explanation of evolution seems to have led him from initially seeking Lyellian-Herschelean foundations for his hypothesis of natural selection toward the less empirical philosophical stance of Whewell, with its reliance on the notion of "consilience." Hodge and Ruse disagree about this shift, however. Hodge argues that Darwin remained faithful to Lyell and Herschel, though he does seem willing to admit, citing the parallels between Darwinian natural selection and wave optics, that this looks more like Whewellian consilience. Ruse thinks he has a very sound case for Darwin's reliance on consilience, again citing the optics parallel. And on more strict empiricist grounds of the type favored by Mill, as Hull points out, Darwin's account ultimately was a "failure." Relatedly, Ruse concludes his article suggesting that the Darwinian approach was only resuscitated in the 1920s after the physical sciences had moved beyond "extreme empiricism," having been forced to deal with the more theoretical developments in relativity and quantum theory. This issue of testability of evolutionary mechanisms arises again in the early twentieth century in the debate between the naturalists and the mutation theorists that Magnus discusses in great depth. It also reappears in Lewontin's essay, which recapitulates that nineteenth-century debate in a very modern major (and) general way.

Lewontin almost tells us that contemporary population genetics, when it seeks empirical validity, is blind, and when it seeks theoretical precision, is empty. This extreme interpretation is countered by some population genetics research programs that do have empirical support for their causal inputs, but at the price of a highly restricted explanatory scope. Lewontin writes, "It cannot be the task of population genetics to fill in the particular quantitative values in the basic structure that will provide a correct and testable detailed explanation of what has happened in any arbitrary case. . . . we lack the necessary observational power and, in practice, will always lack it." He distinguishes between two extreme research programs in population genetics: one a "maximal inferential program meant to give a correct biological explanation of any and all observed evolutionary variation," and the other "a minimally deductive program that provides rules for recognizing acceptable and unacceptable explanations, without reference to any particular observed case." Lewontin adds, "In the end, this minimal program may be the only one that is definitively satisfied."

This is a strong criticism and one that is sure to be controversial both in evolutionary studies and in the philosophy of biology. Lewontin argues for it by citing a broad range of different population genetics research programs and critiquing them incisively. If I understand him correctly, we share similar views about the theoretical significance and the empirical weakness in this discipline (see Schaffner 1993: 89–97, 339–360), but Lewontin, as a major contributor to population genetics, offers much more persuasive evidence than I have seen before for the explanatory problematics in this area.

The two extremes of the evolutionary genetics approaches that Lewontin discusses may possibly be pursuing different epistemic virtues, along the lines of Magnus' analysis. For Magnus, the *experimentalists* had to limit the kind of evidence they were willing to accept in order to attain replicability. The *naturalists*, on the other hand, stressed the virtue of consilience, and yielded on replicability. But consilience can also be employed to rather different ends, in the analysis of experiments, and in a way that prioritizes experimental results, as discussed in Bechtel's chapter.

Bechtel sketches the complex ways that two instruments, the electron microscope and positron emission tomography (PET) scans, attempt to provide empirical, and more precisely visual, information. Use of these instruments involved highly technical problems that needed to be resolved prior to their generating useful visual representations of subcellular structures and mental/brain activities. The key to the instruments' success involved a methodological approach that resonates with Hacking's atheoretical stance regarding visual displays, which Bechtel summarizes in the epigraph to his paper. Bechtel's further analysis is, interestingly, based on consilience, but a consilience that is indirect and analogical. It is thus an ironic twist that this weaker "consilience" approach to evidence, championed by the rationalist Whewell, has become a part of the foundations of what has been called the New Experimentalism, summarized and critically reviewed by Marga Vicedo. But this new consilience approach involves two further ironies: (1) it is less direct, relying on weak analogies, than was Whewellian (and perhaps Darwinian) consilience, and (2) it supports the validity of both *sense* experience and the existence of theoretical *entities*, but not a theoretical structure. Whether the New Experimentalism can flourish based on this fairly mysterious and limited foundation is under vigorous debate. In her essay, Vicedo provides several arguments for an important role for more global theories than the experimentalists seem to wish to admit. In particular, she notes the need to appeal to a rich tapestry of theories, experiments, instruments, and practices to account for Castle's research on the genetics of hooded rats.

BIOLOGY AND A DIVERSE WORLD: VIVA LES DIFFERENCES!

A number of the chapters in this book urge us to eschew the earlier logical positivistic unity of science approach, based on physics as the paradigm science, and to take the diversity of method found in biology as a serious alternative paradigm. George Gaylord Simpson (1964: 107) and Ernst Mayr (1988: 21) had earlier argued for similar theses. In the present volume, Larry Holmes proposes that we take biology as formative, writing that "the nature of biological experimentation cannot be treated only as an extension of methods founded in the physical sciences. The life sciences must be seen instead as a formative site for the origin and development of experimentation in general."

Holmes is primarily interested in the discovery process in biology, and particularly focuses on Krebs' work on the ornithine and critic acid cycles. He notes that although most philosophers do not believe in a "logic of discovery," some have championed it. (He cites Nickels 1980 volumes and my own 1993 book, but does not refer to Darden's more recent 1991 monograph). Holmes speculates that perhaps what I have termed "middle-range theories," which are quite prominent in biology, might permit a logic of discovery because they lack universality. Holmes argues that a fine-structure analysis of Krebs' notebooks demonstrates such a logic, and that Krebs' accomplishments clearly represent important groundbreaking (not Kuhnian "normal") science. I don't think that such a logic works because of the non-universality of the theories reasoned to, because such a reasoning process would still involve ampliative inference. However, the analysis Holmes presents may point toward some type of a non-truth-preserving logic that has yet to be fully articulated or understood. (For some possibilities along these lines, see Darden 1991; Schaffner 1993, chapter 2; Thagard 1988, and also Dunbar 1993.) It is my sense that as philosophers and science studies researchers become familiar with the intricate details of laboratory work, there will be a greater appreciation of the highly rational nature of theory-generating reasoning; and I think it is this intuition that is guiding Holmes.

The idea that biology might point us toward a different type of theory – or a different notion of theory – than that rooted in the physical sciences can also be found in the concluding section of Richards's essay. After his long historical analysis of Darwin's theory's progressivity and recapitulation, and extensive critiques of a number of contemporary commentators on that subject, Richards writes that he has "tried to show the disadvantages of regarding such theories as timeless, abstract entities," and urges that we take them to be "historical entities." This seems right to me, though Richards does not offer any structure as to what exactly this difference might amount to. I suspect that Richards might find some philosophical attempts to provide such structure useful, many of them growing out of Kuhn's challenge in his seminal *Structure of Scientific Revolutions* (1962, 1970). In fact, since Kuhn, many philosophers have taken theories to be evolving entities, ranging from Toulmin (1971), Lakatos (1970), and Laudan (1977), to Kitcher (1989, 1993), Hull (1988), and even my own approach to what I term a "temporally extended theory" (1993, esp. chapter 5).

Maintaining that there is any single type of theory that predominates in biology is likely to fail, and I myself have suggested that we are likely to find other types of theory than what I have mentioned here as "middle-range theories" (for a specific suggestion that Wright-like evolutionary theories and models of protein synthesis should be distinguished, see Schaffner 1993, chapter 3). Helen Longino also urges that we avoid a search for unification in her arguments for substantive and methodological pluralism. She provides us with sketches of five different areas in biology that support a pluralistic approach, and my suspicion is that biologists, and those engaged in social, historical, and philosophical studies of biology, will have to adopt a diversity of kinds of analyses as both confront nature in all its complexity. Two essays in the present volume address the subject of developmental biology, and I have saved my comments on these for last. The history, philosophy, and social studies of biology have seen fashions come and go in an overlapping manner. A focus on evolutionary theories has led the modern period, but that has been at least equalled, if not supplanted, by work in molecular genetics and more recently in neuroscience. It could be argued that developmental biology is "the next big thing." This scenario is driven both by breakthroughs in developmental genetics and by synergy between evolution and development (or "evo-devo") studies (Roush and Pennisi 1997). In addition, philosophers who were once occupied with the issues of reductionism and antireductionism appear to be triangulating on some common ground in the developmental biology area (see Culp and Kitcher 1989; Kitcher 1998; Rosenberg 1997; Griffiths and Knight 1998; Schaffner 1998a,b; Wimsatt 1998), though vigorous differences continue to exist.

In her chapter in this volume, Fox Keller writes that "nowhere has the goal of a causal account remained more elusive than in the core question of developmental biology, namely, how does a zygote develop into a multicelled organism. She adds that "more than any other biological phenomenon, it is the process of embryogenesis that has led so many thinkers of the past to conclude that biology requires distinctive modes of explanation." Jane Maienschein, in her chapter, covers three or four critical debates in developmental biology and shows the extent to which this complex area was ripe for diverse epistemological positions that importantly conditioned the work of Wolff, Bonnet, Roux, Driesch, Wilson, Golgi, and Ramon y Cajal. Interestingly, she notes that T. H. Morgan, initially an embryologist, turned to his stunningly successful research program in *Drosophila* genetics because of the clarity and forcefulness of experimental results in the genetics area, in contrast to developmental biology at that time.

Fox Keller directs most of her attention to the current state of developmental biology. She critiques the genetic cause approach to developmental biology, though she includes as background to her account a historical review of Waddington's epigenetic program. From her perspective, the genetic paradigm must be juxtaposed to an alternative perspective that she favors – "one that takes as its starting point not genes, but rather, the manifest robustness of the developmental process."

Fox Keller's alternative perspective focuses on the perplexing phenomena of biological redundancy, and highlights the failure of the one mutation – one phenotypic change approach. She cites the words of Nobel laureate J. Michael Bishop, who has stated that we must move beyond simple switch models to address "an elaborate network with hundreds (if not thousands) of nodal points," and who has called for a more "global view of the molecular circuitry." Keller notes that the nature of an explanation in developmental biology is by no means obvious, but she appears convinced that it will not be a simple gene-based form of explanation. In this view she is joined by developmental systems theorists such as Oyama (1985), Weele (1995), and Griffiths and Gray (1994), and also by Lewontin in his writings on developmental biology (1993, 1995). The extent to which a global non-gene-based analysis is defensible is a controversial question (see my 1998a,b for references), but it strikes me that developmental biology will provide what is perhaps *the* field in which traditional questions about biological explanation and epistemology receive new interpretations and a new urgency.

CONCLUSION

The contributions to this volume demonstrate that epistemological issues are tightly interwoven with scientific inquiries in the biological sciences, both historically and into the present day. To this author, it is the recurrent themes of empiricism in both its more classical and its newer forms, the roles and interpretations of experiments, and the complexity of explanations that are most exciting. Historians, philosophers, social scientists, and biologists will be debating these themes well into the next millennium. A close reading of this volume should prepare students of methodology well, both in surveying some of the most salient issues and in anticipating the vigorous exchanges to come.

REFERENCES

Culp, S. and P. Kitcher. 1989. "Theory Structure and Theory Change in Contemporary Molecular Biology." *British Journal for the Philosophy of Science* 40: 459–483.

Darden, L. 1991. *Theory Change in Science: Strategies from Mendelian Genetics.* New York: Oxford University Press.

Dunbar, K. 1993. "Concept Discovery in a Scientific Domain." *Cognitive Science* 17: 397–434.

Griffiths, P. and R. Gray. 1994. "Developmental Systems and Evolutionary Explanation," *Journal of Philosophy* 41: 277–304.

Griffiths, Paul E. and R. D. Knight. 1998. "What Is the Developmentalist Challenge?" *Philosophy of Science* 65: 253–258.

Hull, D. 1988. *Science as a Process: An Evolutionary Account of the Social and Conceptual Development of Science.* Chicago: University of Chicago Press.

Kitcher, P. 1989. "Explanatory Unification and the Causal Structure of the World." In P. Kitcher and S. Salmon, eds., *Scientific Explanation,* 410–505. Minneapolis: University of Minnesota Press.

Kitcher, P. 1998. "Battling the Undead: How (and How Not) to Resist Genetic Determinism." To appear in a festschrift for Richard Lewontin. Privately circulated ms.

Kuhn, T. S. 1970 [1962]. *The Structure of Scientific Revolutions,* 2d ed. Chicago: University of Chicago Press.

Lakatos, I. 1970. "Falsification and the Methodology of Scientific Research Programmes." In I. Lakatos and A. Musgrave, eds., *Criticism and the Growth of Knowledge,* 91–196. Cambridge: Cambridge University Press.

Laudan, L. 1977. *Progress and Its Problems*. Berkeley: University of California Press.

Lewontin, R. 1993. *Biology as Ideology: The Doctrine of DNA*. New York: HarperCollins.

Lewontin, R. 1995. *Human Diversity*. New York: Scientific American Library.

Mayr, E. 1988. *Toward a New Philosophy of Biology*. Cambridge: Harvard University Press.

Nickles, T., ed. 1980a. *Scientific Discovery, Logic, and Rationality*. Dordrecht: Reidel.

Nickles, T., ed. 1980b. *Scientific Discovery: Case Studies*. Dordrecht: Reidel.

Oyama, S. 1985. *The Ontogeny of Information: Developmental Systems and Evolution*. Cambridge: Cambridge University Press.

Rosenberg, A. 1997. "Reductionism Redux: Computing the Embryo." *Biology and Philosophy* 12: 445–470.

Roush, W. and E. Pennisi. 1997. "Growing Pains: Evo-Devo Researchers Straddle Cultures." *Science* 277: 38–39.

Schaffner, K. F. *Discovery and Explanation in Biology and Medicine*. Chicago: University of Chicago Press, 1993.

Schaffner, K. F. "Genes, Behavior, and Developmental Emergentism: One Process, Indivisible?" *Philosophy of Science* 65: 209–252.

Schaffner, K. F. "Model Organisms and Behavioral Genetics: A Rejoinder." *Philosophy of Science* 65: 276–288.

Simpson, G. G. 1964. *This View of Life*. New York: Harcourt, Brace & World.

Thagard, P. 1988. *Computational Philosophy of Science*. Cambridge, MA: MIT Press.

Toulmin, S. 1972. *Human Understanding*, vol. 1. Princeton, N.J.: Princeton University Press.

Wimsatt, William C. 1998. "Simple Systems and Phylogenetic Diversity." *Philosophy of Science* 65: 267–275.

Weele, Cor van der. 1995. *Images of Development: Environmental Causes in Ontogeny*. The Hague: CIP–Gegevens Koninklijke Bibliotheek.

5/2000

University of St. Francis Library

3 0301 00207291 2

Biology and Epistemology

This set of original essays by some of the best names in philosophy of science explores a range of diverse issues at the intersection of biology and epistemology. It asks whether the study of life requires a special biological approach to knowledge and concludes that it does not. The studies, taken together, help to develop and deepen our understanding of how biology works and what counts as warranted knowledge and as legitimate approaches to the study of life.

The volume is organized into three sections. The first focuses on questions surrounding a central idea of the nineteenth century: evolution and its contemporary philosophy of science. What view did Darwin and the leading evolutionists hold concerning the nature of evidence and its relation to theory? What was the relation of philosophy of science to biology? The second set of papers moves to this century and to the virtual explosion of laboratory and experimental research. The papers here explore the nature and use of evidence, considering such central questions as what is data, when data counts as evidence, and what role experimentation plays in revealing knowledge about living nature. The third section discusses the nature and role of argument, considers issues of objectivity and other goals in science, and examines the way those have changed over time in response to a diversity of factors.

This is an impressive team of authors, bringing together some of the most distinguished philosophers of science today. The volume will interest professionals and graduate students in biology and in the history and philosophy of science.

Richard Creath is Professor of Philosophy at Arizona State University. Jane Maienschein is Professor of Philosophy and Biology at Arizona State University.